ECOLOGICAL POLITICS IN AN AGE OF RISK

For
Etty Hillesum,
whose diaries accompanied me
in the writing of this book

According to a Red Cross report, Etty Hillesum was
killed in Auschwitz on 30 November 1943; her
parents and brothers also lost their lives there.

Ecological Politics in an Age of Risk

ULRICH BECK

Translated by Amos Weisz

Polity Press

English translation © Polity Press 1995
First published in Germany as *Genengifte: Die organisierte Unverantwortlichkeit*
Copyright © Suhrkamp Verlag Frankfurt am Main 1988

This translation first published in 1995 by Polity Press in association with Blackwell Publishers.

Reprinted 2002

Published with the financial support of Inter Nationes, Bonn

Editorial office:
Polity Press
65 Bridge Street
Cambridge CB2 1UR, UK

Published in the USA by
Blackwell Publishers Inc
350 Main Street
Malden, MA 02148, USA

Marketing and production:
Blackwell Publishers
108 Cowley Road
Oxford OX4 1JF, UK

ISBN 0 7456 0763 2
ISBN 0 7456 1377 2 (pbk)

A CIP catalogue record for this book is available from the British Library and the Library of Congress.

Typeset in 10½ on 12pt Sabon
by Best-set Typesetter Ltd, Hong Kong
Printed in Great Britain by MPG Books Ltd, Bodmin, Cornwall

This book is printed on acid-free paper.

Contents

Translator's Note

Several problems face the translator of this book. First, and most obvious, is the fact that it was written before the removal of the Iron Curtain in 1989, and hence some anachronisms will strike the reader. Second, English frequently offers several words where only one is available to the German-speaker and -writer. Thus *Sicherheit* can be translated as 'security', 'safety' or 'certainty', depending upon the context. I have tried to give the appropriate English word, at the cost of a loss of linguistic word-play. Thus 'reactor safety' (*Reaktorsicherheit*) is mentioned in the context of the 'security state' (*Sicherheitsstaat*). Similarly, *Politik* can be translated as either 'politics' or 'policy'; and so forth. Third, and last, a direct translation of the cadences of Ulrich Beck's prose would appear impossibly involuted, at times, to the English eye. As far as I could, I have shortened some of the longer sentences, sometimes changing the order of clauses.

Amos Weisz

Preface

The thoughts expressed here are the fruit of many conversations and discussions. Above all, I have woven into the fabric of the book ideas that I owe to my life with my wife and colleague Elisabeth Beck-Gernsheim. Peter Berger, Wolfgang Bonss, Ronald Hitzler, Christoph Lau, Maria Rerrich, Renate Schütz and Rainer Wolf made trenchant comments and gave helpful advice on the book's structure. I also thank Reiner Keller, Gerhard Mutz and Claudia Wurst for giving greater precision to some of my formulations.

Many people have struggled through the false turns of the early drafts and given unstinting advice from which subsequent readers will benefit, especially Peter Gross, Jobst Günther, Heinz Hartmann and Dieter Mertens. Angelika Schacht and Gerlinde Müller have also been endlessly reliable both in typing the text and in making sure that I was left time in which I was free to write.

I thank the Deutsche Forschungsgemeinschaft (German Research Council) for their generous support.

Lastly I acknowledge with gratitude the contribution of the gleaming Starnberger See.

Ulrich Beck

Introduction: *The Immortality of Industrial Society and the Contents of this Book*

The theme of this book is the paradigm confusion involved in the management of hazards. The challenges of the atomic, chemical and genetic age at the turn of the twenty-first century are discussed in conceptual and prescriptive terms that derive from the early industrial society of the nineteenth and early twentieth centuries. A multiple disjunction separates the risks of early industrialization from the hazards of technologically advanced civilisation.

1   These latter hazards cannot be delimited spatially, temporally, or socially; they encompass nation-states, military alliances and all social classes, and, by their very nature, present wholly new kinds of challenge to the institutions designed for their control.

2   The established rules of attribution and liability – causality and guilt – break down. That means that their careful application to research and jurisdiction has the contrary effect: the hazards increase and their anonymization is legitimated.

3   The hazards can only be minimized by technological means, never ruled out. In an age of worldwide growth of large-scale technological systems, the least likely event will occur in the long run. The technocracy of hazard squirms in the thumbscrews of the safety guarantees which it is forced to impose on itself, and tighten time and again in the mass-media spotlight of the bureaucratic welfare state.

4   The lack of provision for catastrophe plainly exposes the paradigm error, the bewitchment of reason caused by the false belief that the twentieth century is only the continuation of the nineteenth. If the rafters are on fire, the fire brigade will arrive, the insurance company will pay, the necessary medical attention and so forth will be given.

This security system, which anticipates social provision for the worst conceivable case, broke down with the advent of large-scale (nuclear, chemical, ecological, genetic) hazards. Accidents now frequently cause irreversible damage and destruction that may have a determinable beginning but no foreseeable end. Yet it is not only 'accident' statistics that fail to address the historically unprecedented fact of artificial disasters of undeterminable extent; the guiding idea of economic compensation, which has prevailed hitherto, also fails to meet the case.

These open-ended, ultimately irremediable large-scale hazards are, however, forced upon the heightened safety consciousness of citizens with every means at the disposal of state authority. In so far as the paradigm confusion, upon which industry and politics have built their safety guarantees, is revealed in the sequence of disasters, near-disasters or hushed-up disasters, a great deal happens, even if it does not appear to. The social explosiveness of hazards develops its own political momentum: risks consciously taken must be socially answered for, as they endanger the lives of everyone and stand in open contradiction to the state's institutionalized pledges of safety and welfare.

It is not only seals in the North and Baltic Seas that suffer agonizing deaths. Chemicals that are today an integral part of the civilized world have arrived in profusion at the penguin colonies of the Antarctic. Yet the law is circumscribed by the unquestioned assumptions of a different epoch; it can intervene only when the 'sole culprit', that vestige of tradition, has been apprehended in the world of chemicals. In the legalized international traffic in harmful and toxic substances, the sole culprit is also an extinct species. As long as the universal dissemination of poisons is ensured by the absence or laxity of maximum pollution levels, holding a single individual liable is comparable to trying to drain the ocean with a sieve. This is precisely what organized irresponsibility means. The interpretation of the principle of causation in individual terms, which is the legal foundation for hazard aversion, protects the perpetrators it is supposed to bring to book. It is absurd how an ostensibly protective judicial system, with all its laws and bureaucratic pretensions, almost perfectly transforms collective guilt into general acquittal.

Safety issues that convulse societies from the Urals to the Atlantic are, in the final analysis, illegitimately decided by the corporation of engineers of our high-risk civilization. These decisions are taken under cover of the empowerment formula, 'state of science and technology', which puts the essence of safety laws, namely their small print, into the hands of corporate owners and experts. It is as if safety experts could claim certainty for judgements which are always and necessarily based on probability; as if engineers or physicians, however brilliant in their own field, knew anything about the political explosive they undertake to guard closely behind safety barriers, which are in turn constructed out of highly permeable probability calculations; and as if, in an age that considers itself democratic, they were entitled to sit on the throne and bid us to live dangerously.

In the absence of societal debate and extra-parliamentary opposition, manipulation of the genetic code continues apace, supplanting the cultural invariants of life as it is known. Governments need to be shocked by newspaper reports into asking what society is in the test-tubes of the genetic engineers, and to which (now biologically manipulable) laws it is subject. People are alarmed. Yet while the concerns of 'progress' are free of public control, this vague sense of alarm can find no point of application. By virtue of the social structure there is no site, no obstacle, no decision, no decision-maker, that allows for dissent or assent in the maze of 'progress'. There are only extreme, and extremely one-sided, burdens of proof, thrust upon anyone who registers misgivings.

How is it possible that our society fails to recognize the vast challenges it faces. All past societies believed, always falsely, in their own immortality – while we Olympians of today have truly scaled the peak of development. Indeed, this is precisely what distinguishes our epoch from all others, none of which thought any differently. *Post-histoire*, the illusion of having reached the terminus of the history of societies, is in truth the most universally valid law of thought in history. The provincial self-consciousness of the age, its incapacity to look beyond the narrow horizon of the prevalent unquestioned assumptions, was and is the end-of-societal-history thesis.

Thus in the dim and distant past, people were commanded by empty stomachs and by custom to keep moving, in order to hunt bears and gather berries. At night, when the wood had again failed to catch fire, they may have mused: perhaps the alternative, sedentary life is feasible after all, and desirable. Then came the knock-down refutation: how is a settled person (who knows the nomadic way of life only from package tours) to feed himself once the bear has been bagged, and all the berries have been picked?

Hardly less convincing are proofs from agrarian societies that the hierarchy, power, inequality, poverty and splendour of the feudal order are the only possible form of life: where human beings become lords, peasants or serfs by virtue of birth, and thereby through God's intercession, the social order is natural and therefore good (just as free-range eggs are better than battery-farmed eggs; or was that true only until the 'human failure' at Chernobyl, and untrue since?). The unquestioned assumption that one feudal society is always replaced by another is thus founded upon nothing less than the immutability of nature. To assume the contrary would be comparable to trying to leap out of the window and fly upwards.

For industrial society, the unquestioned assumption that every industrial society is succeeded by another is even more obvious. Our epoch distinguishes itself from all others in having replaced the principle of constancy by that of change. Since everything is in constant flux, the process is always more or less the same. After every industrial revolution, which turns upside down the conditions of industrial society as we have known it, the familiar forms emerge anew – classes or strata, competition for world markets; and, fresh as ever, the welfare state, the scientific attitude, the family, waged labour, the professions, businesses, industries, etc., with men's and women's roles perhaps losing a bit of their sparkle. That is, we have a social system which can perhaps be distinguished with scientific precision from its predecessor by its slightly higher level of industrialization.

The final and real reason for the immortality of industrial society, the one that will be examined here in detail, can be seen from the fact that now, in its current late phase, it has at its disposal, and has begun to utilize, the earth's mortality, together with everything that crawls on or flies above it. Our epoch has taken progress so far that a minimal exertion may relieve everyone of all further exertions. Ours is the age of the smallest possible cause for the greatest possible destruction. In accordance with the law of intransience of conceptual epochs, our era and its society have achieved and proved beyond all doubt the immortality of its way of living, thinking, working and running affairs; of its scientific, political and legal practices.

We have done away with life after death, and placed life itself under permanent threat of extinction. Nothing could be more transient. Yet we have done more: we have elevated transience to a principle of progress, released the potential for self-destruction from its restriction to warfare, and turned it, in manifold forms, into the norm: failsafe and ever more failsafe atomic power plants; creeping and galloping pollution; the latest creations of genetic engineering, and so forth. That is how we live at the summit of world history, where the future spreads out over the plain of the nothing new. More! Bigger! Keep it up!

The future of industrial society is industrial society, and the future of that is again industrial society. Just as the future of hunter-gatherers is, was and always will be hunter-gatherer society, and likewise for the future of feudal society.

If this book nonetheless rebels against that iron law, the law of the eternal insuperability of the prevalent conceptual epoch in all human history, and does so solely by appealing to a human understanding distanced by the practice of sociology, I shall be taking upon myself an intolerable burden of proof. How is it possible to champion and vindicate historical appearance, of all things the most ridiculous and ragged of all excuses, against a hydra-headed social science armed to the teeth with expensive theories and figures? It is utterly impossible, and should therefore be held in this book's favour as a first mode of self-refutation.

For the record: whoever takes my arguments on board does so in spite of my own misgivings, and therefore on their own initiative and according to their own lights.

A major legacy of the industrial-capitalist colossus is the unbroken dominance of the false alternative: whoever disputes the rationality of science – so it is claimed – awakens the slumbering ghosts of irrationality. In the debate over the Enlightenment, one side defends the idea that the past has a future, while the other proclaims the end of the Enlightenment. In the name of what, or of whom this is done remains nebulous. Everywhere the same alternative is offered: either modernity or postmodernity. For or against. Yet even those who dispute the proposition hypostatize it into a constant, monolithic block. The idea of neither modernity nor postmodernity, however, the reality of the excluded middle, remains as alien as if it had fallen from another star.

The ensemble of identifications – industry = progress = science = enlightenment = modernity – is now in motion, with a continuity and momentum that determine industrialism's law of development. Therefore, neither these equations nor the positions of their critics are valid any longer. Even the most radical opposition – the condemnation of all that once meant and promised the triumph of reason, rationality, comprehension – is rendered conventional because it does not notice that its other, and hence its own self, has been deprived of co-ordinates. If this appraisal is correct, then social analysis must start afresh from its foundations, and on its methods of diagnosing the age. 'Some time or another', wrote Max Weber at the beginning of this century, 'the colour changes: the meaning of unreflectively adopted viewpoints becomes uncertain, the path is enveloped by darkness. The light of the great problems of culture has moved on. Then science too prepares to change its position and conceptual apparatus' (1968, p. 214). Max Weber's 'some time or another' is our present day, our aporia and our project.

The now false dichotomy between nature and society is at the heart of the first part of this book, which takes for its theme the aberrations to which speech and praxis in terms of the nature/society dichotomy lead today, when destruction and protest point to a common stratum, as yet unperceived, in nature and society. The result, as practically applied, is dead ends, variants of a systematically fomented fatalism, a fatalism of (post)modernity.

Antidotes, which are sought in the second part of the book, become discernible in the maze of false alternatives when what appears in the guise of 'natural destruction' is revealed as a social relation – objectified errors of naive industrialization, whose cultural sanction is being revoked; the threat to the existence of markets, industries and regions; the avoidable consequences of the organized non-liability that industrialism has become over the centuries.

To continue speaking of 'risks' in the case of reproductive medicine, and particularly of human gene technology, would be an anachronism, as chapter 1 demonstrates. The genetic code represents a unique field of operations. Repercussions, mistakes that develop here, change the biological constitution of living things, usually irreversibly. To that extent they can neither be treated as anonymous nor blamed on the 'environment'. The product itself is life – or quite the reverse.

Biologists and physicians are smuggling in a new age, swaddled in 'normality', beyond the limits of the acceptable. They seek shelter from unpleasant questions in giddy heights of abstraction. For example, they draw an analogy with cheese manufacture through the centuries, in order to establish a connection with preparations for rewriting the genetic text. Human nature, nature *tout court*, is becoming malleable beyond the limits of natural kinds. In the continuation of the Enlightenment to technological ends, the relationship between subjectivity and nature, between subjectivity and society, is placed at the disposition of society. In principle, subjectivity and society are becoming 'plastic' (van den Daele) – directly and without the intervention of executive or legislature, without the baffling profusion of judgements or conflicting interpretations – and sentence is passed in the sterilized biological and chemical laboratories.

Neither looking away nor cheering will help us: the successes of reproductive medicine and of (human) genetic engineering are bestowing upon us a eugenic age. Chapter 1 examines the possibility, now becoming a reality, of modernization reverting to barbarism. It explores the nightmare of my generation, the children of those who perpetrated and tolerated the Nazi terror, that its actions and omissions will once again, in other forms and by other means, turn madness into normality.

The 'natural world', sapped by society and industrially endangered, has become the battleground for its own survival, as described in chapter 2. Yet the ecological movement remains trapped in a naturalistic misunderstanding. It reacts to and acts upon a blend of nature and society that remains uncomprehended, in the name of a nature no longer extant, which is at the same time supposed to serve as a model for the reorganization of an 'ecological society'.

This confusion of nature and society obscures from view another central political insight: the independence of destruction and protest. Protests against the despoliation of nature are culturally and symbolically mediated. They cannot be deciphered according to the calculus of hazards, for instance, as diagnosed by natural science, but must be interpreted through the inner and personal experience of social ways of life.

Naive naturalism and the technology of hazard hold everyday life, politics and the protest movement under their spell. It is the thesis of chapter 3 that they allow the establishment of the prevailing, extremely unequal burdens of proof and let the current, historically inapplicable rules of attribution go unchallenged. Large-scale hazards are not hazards-in-themselves, clearly to be grasped and delimited from normality on the strength of technological-medical authority. Rather, they are the concern of all, and in a new way. Manifold policies, cultural assumptions, mechanisms and rules are built into them: maximum pollution levels, rules of attribution, principles of compensation, acceptance, etc. To ignore this fact is to lose one's way in the labyrinth of provable unprovability that science and law have become, in their ahistoricism and incorrigible abstraction.

Once again, the canon of sociological classics blocks one's path at the outset of the search for antidotes in part II: Karl Marx's theory of capitalist exploitation and Max Weber's cage of bureaucratic subordination are only two milestones along the path to the dead end to which sociological thinking, with its excessive partiality to institutional objectivity, has condemned action. The tradition of intervention and resistance has wasted away, and is decaying into such conceptual ruins as 'class struggle', 'revolutionary subject', 'subjective factor', 'critical public opinion', a list which could be extended much further.

The problem of politically deriving a 'lever' for change is then resolved in chapter 4, paradoxically as it may appear at first. The real, most influential adversaries of the nuclear power industry, for instance, are not social movements, campaigning journalists or dissenting experts. These are all indispensable, and their role in the past decade's ecological revolution of consciousness need not be minimized in any way to see that the most convincing long-term opponent of the nuclear power industry is the nuclear power industry itself.

Even if the institutions of hazard production and administration reign supreme and their 'symbolic detoxification policy' proves effective; even if social protest abates and its political scope remains circumscribed; all this can be shown, no less realistically, to be offset by the objective counterforce of hazard. The latter is constant, enduring, not bound to interpretations that deny it, and present even where the demonstrators have long since tired. The probability of improbable accidents grows with time and with the number of completed major technological systems; every 'incident' awakens memories of all the others, everywhere in the world. The world has become a testing ground for risky technologies, and thus also a potential refutation of the safety guarantees of state, economic and technical authority.

It takes only the suspicion of a catastrophe to bring about change in the security-state system of organized non-liability: the danger of annihilation dismantles the basic consensus that has until now put up with the internal and external conflicts of individual and common interests. The 'invisible hand' turns into an 'invisible saboteur', which cannot, or can only barely, be apprehended, and thus as it were 'covered', by the current categories of legal and scientific hazard assessment. Bertolt Brecht quipped that the sole difference between a bank director and a bank robber is that one steals people's money legally and continuously, while the other robs illegally and at intervals. This notion now applies to the incomparably graver case of the threat to life.

Large-scale hazards can thus be interpreted sociologically as a kind of revolution, become independent, which the conditions have instigated against themselves. The industrial dynamic finds its immanent 'adversary' in the virtually autonomous disclosure of hazards, depending upon (a) insidious, suspected disasters, (b) cultural sensitization, (c) the attentions of the mass media and (d) the resulting divisions and conflicts in the economic camp between those who profit and those who lose by the risks.

This political theory of major technological-ecological hazards is developed in chapter 4 as distinct from two positions – 'scientific objectivism about hazards' and 'cultural relativism about hazards'. The sociological objectivity of the concept of hazard proposed here is hence predicated not on technical alarms, but upon the institutionalized safety and control guarantees of the developed welfare state, which enter into contradiction with the bureaucratically perfected legalization of hazards. Policy, law and government have internalized the safety constructs of industry and of research technology, and are now squandering their authority as the error of the century within the technology-centred philosophy that guides them becomes ever more apparent. This estimation of a concealed, responsive self-politicization of hazards in public perception, politics and the hazard bureaucracy is worked out theoretically in chapter 4, with reference to Max Weber ('Purposive rationality and the rationality of risks') and François Ewald, who depicted the emergence and self-endangering of the 'assurance state'.

Like the hazards themselves, the social upheavals that result from their suppression and consequent outbreaks can no longer be de-limited either socially or temporally. Science, and particularly tech-nology, is only one of the areas where conflicts over progress erupt. For risks, which must now be calculated according to all the rules of the art, are a form of involuntary self-refutation of scientific rational-ity – as is shown in chapter 5. Not only is science internally divided, continually contradicting its own safety claims, but advances in the science of risk represent a decline of scientific authority on safety matters. Also, a science that extends its claims of accuracy to the investigation of repercussions, turns in fact into a theatre of the absurd: precision refutes precision. Risk calculations can be variously interpreted, and so they return full of mathematics and contradictory recommendations. These are supposed to manufacture acceptance, yet remain dependent on it. Maximum acceptable levels have to be fed into the calculations from which they are supposedly deducible.

Ultimately, danger, no longer subject to experimental logic, turns even that on its head: for nuclear power plants to be examined for safety, they must first be constructed. The application precedes the examination. The precondition for investigating their safety is that it will be confirmed. What this has to do with good old natural science is a ticklish question. In the field of large-scale hazards, the thoughts and deeds of technology and the natural sciences belong to distinct eras. Not its deeds, but only the representations of its deeds, can (perhaps) be justified by the canon of rules they call science.

The system of the economy (chapter 6) also gets ensnared and politicized in the contradictions of organized non-liability. Only in appearance, and for the producers of risk themselves, is the environ-ment merely environment. From another point of view, in socialized nature the 'environment' is the economic basis of those industries and regions that live off the commodification of nature: fisheries, the food sector, holiday resorts and tourist destinations, and also the trade sector and consumer goods industries. As the floodgates of poison open (through the absence of maximum pollution levels, inapplicable principles of causal attribution and juridical fictions), an explosive political situation emerges. In the omnipresence of harmful substances, a spark of information ignited by the mass media can destroy whole markets and industries. The victims cannot be specified or determined in advance. Where despoliation is unattributable, the economy, the public and the media begin to play Russian roulette under cover of the category of 'environmental' hazards deriving from a different age.

For all that the outcome is uncertain, the chances of being affected are very unequally distributed: this time, the 'proletariat' of risk society includes not only various kinds of worker, but also promising branches of enterprise, possibly even whole regions (states on the North and Baltic Seas, industrially undeveloped woodland regions). These have to pay with their economic lives for the legalized, total pollution that systemic unattributability conceals.

Here there are clear differences from the old class conflicts over the distribution of social wealth: if in those days labour and capital stood opposed to one another (and still do in this respect), the battle to distribute away the 'poisoned cake' turns capital against capital – and, consequently, occupational group against occupational group. Some industries and regions profit by this, others lose. But a key question in the struggle for economic survival has become how to win and exercise power, in order to foist on others the consequences of social definitions of risk.

There was a time, in the entrepreneurial paradise of early capitalism, when industry could begin projects without submitting itself to controls and agreements. Then came the era of state intervention, when this was possible only in consultation, and on a foundation of laws and regulations. Today even this will no longer suffice. Such arrangements can be negotiated and signed, but company managements feel exposed to further conflict, resistance, public denunciation and suspicion. These not only call into question the agreements reached on the basis of law, but exert unforeseeable and incalculable control even over the details (from waste disposal, through the material composition of products, to the details of manufacture) that were formerly the monopoly of technology and management. The defenders of the old order console and persuade one another that this is 'irrational' and 'ideological', a product of mass-media hysteria and long-haired layabouts – symptoms which can be 'cured' at the next recession with the silent whiplash of economic circumstances. That is not so. First, it is far more the expression of a more developed democracy, where an expanded civic consciousness refuses to be excluded from participation without a fight – in making decisions that intrude upon our lives more palpably and hazardously than those susceptible to parliamentary measures. Second, they are simply indications of the range and political potential of industrial hazards. Unlike their early industrial forebears, these dangers are no longer restricted solely to the workplace or to the freedom of consumer choice, but also include the lives of all 'third parties', including generations yet unborn.

Such historic outbreaks of conflict cannot however be routinely packaged, as in the good old days, by means of new technology, then crowned with politically renewed safety pledges or sweetened by this or that law. They can be resolved only through historical learning processes and changes that perceive the secular error, and which this time aim to overcome organized irresponsibility, i.e. the power relations of definition (chapter 7). Ecological devastation and social divisions cannot in the end be wished away be gesture politics, the centralization of data or the creation of new government bodies. They can only be overcome by rules of decision-making that break up and democratize the concentration of power on questions of definition, because the problem of attribution can only be solved in this way. A change in the relations of production was required (through social insurance, rules for participation, union power and workers' parties) for social modalities to emerge that made a regulated conflict possible. Similarly we will need new rules for consultation and decision-making, and a redistribution of the burden of proof – radical changes affecting the foundations of industrial production, as well as those of science, law and politics – to open up the possibility of no longer endangering, along with the environment, health, civic rights and related industries.

History teaches us that concentrations of power cannot be thus dispersed and democratized through questionnaires or by learned appeals to the understanding. It cannot be done without conflicts over progress, which owing to the universality of hazards are no longer restricted to one area, but penetrate every region and level of society. The technocracy of hazard and its advocates must fry in the purgatory of their false safety pledges. Thus no help is to be expected from a kind of 'political acupuncture' (although a politics of multiple precision jabs can be very effective). Nor will some 'revolutionary subject', this time perhaps eugenically improved, drop it into the laps of those who wait. Nor will it result from the ardent hope for reason, discourse and openness, indispensable to be sure, but no more than that. If the analysis presented is correct, then we should not think and act in the opposing categories of politics and citizens, of bureaucracy and social movements. We should put the totality of bureaucratic-industrial-political supremacy, with its immanent division into the heralds and the transgressors of safety standards and life-norms, at the centre of an oppositional politics. This will derive its power not only from within, but from the political adroitness with which it exploits institutionalized political schizophrenia, so that under the prevailing universal imperative to protect life, the contrary practices that endanger and destroy it will be found out and made public. In other words, one must bring out the implications of the insight that the nuclear power or chemical industries etc. are their own most powerful and enduring adversaries. For example, by taking at its word the chemical industry's claims, published in full-page advertisements, to be the very quintessence of concern for humanity, one might bring it to bay by following up its own errors. So that the evangelists of ecological ethics – as our fallen brothers of the chemical industries have to style themselves with the discovery of each new ecological sin – finally provide the criteria and clues that convict them of their sins. It is clear that the rules of the game will have to be changed. How that may be done will be revealed only in the final section, and even there only incompletely.

> At the end of their treatises, in which the inevitable end (of industrialization, civilization, humankind, life on the planet) is convincingly depicted if not proved, they always tack on a chapter in which they stress that there is another way ... which curiously puts into contradiction the appalling prophecy of disaster and the harmless exhortations with which we are let off. This contrast is so glaring that each side of the argument damages the other. At least one of them sounds unbelievable: either the closing sermon that would reassure us, or the analysis that seeks to terrify us. (Enzensberger 1973, p. 32ff)

Many people justly noted and criticized the same imbalance, clearly an occupational hazard of sociology, between the diagnostic analysis and the little chapter of hope at the end of my book, *Risk Society: Towards a New Modernity* (1992). One is naturally reluctant to formulate an answer to the question on everyone's lips: what are we to do? Perhaps, on the contrary, one is too ready to oblige in this learned milieu, in which the problem of a new world order consists principally in its formulation. While higher and higher levels of hazard become the norm, and while safety levels progress ever upwards, our lives continue ever more normally, ever more hazardously. Under these conditions, the question about the meaning of 'What is to be done?' has already been answered.

The paradoxes of this question have split this book and maintained the split between 'Dead Ends' (part I) and 'Antidotes' (part II). This dilemma continues, even though the book's title and the sequence of chapters might appear to indicate the contrary.

The argument of the book can only be as powerful as the reader judges it to be. I have evaluated specific cases and empirical data, where these were available and accessible to me, and articulated alternative theses. Yet a great deal remains speculative. That is not my fault alone, but is also due to the state of research into the social sciences, which have not exactly been keen to pursue the questions thrown up here. To put it bluntly, I am perhaps the least certain participant in the uncertain science with which I deal. The lack of ifs and buts in the formulations is a question of style. Let this fact be taken out of parentheses and writ large once and for all.

Yet the uncertainty of all claims to knowledge, as revealed to consciousness by thorough inquiry, need not end up as pussy-footing. This book also intends to demonstrate that. Anyone who has grasped the fragility of what is most certain can fall silent, turn cynical, get into a rut – or else take the opportunity of transforming prevalent concepts, once having discerned their fallibility. If we are correct in asserting that the self-endangering, 'civilized' world is no more than a (disproved) hypothesis that we have not yet put behind us, now is the time for the counter-hypothesis. The error to which the ossification of scientific concepts leads can only be broken up by an interplay between the internal and the external, with the courage that draws its strength from the will to know.

After the technological and scientific superstition that keeps this age in its thrall, though now under the tyranny of self-destructiveness, perhaps some old-fashioned enlightenment can begin anew. Preparing for this has generated the pleasure, the rage and the profound pessimism that animate this book.

PART I

Dead Ends

1

Barbarism Modernized:
The Eugenic Age

We are at the dawn of another golden age, this time perhaps tinged with green. Since the late 1950s modern biology has solved the mysteries of the cell nucleus. That heralds 'the eighth day of creation', as the euphoric scientists claim (and they must know about these things), and that is why the band has struck up Beethoven's Ninth Symphony. It is still unclear how things will look on the ninth day or the tenth. Once again there is nothing here that is new, unprecedented or undesirable. On this topic I examine first some prospects for the foreseeable future; next, the inequality of burdens of proof in the field of decision-making; then the question of the social consequences; and lastly that of the 'modernization of eugenics' resulting from human genetics.

The technology of creation: humans and nature off the drawing board?

At the top of the list of biology's promises is the liberation of humanity from the nightmares that haunt us. Our agriculture is being rearmed for the battle against famine, particularly in the Third World. The plan is to let the wine lakes, milk lakes and meat mountains grow large enough somehow to benefit (there is a slight problem here, but surely it can be solved?) the poorhouses of the world.

Miracle plants are to bring this about. They stand the folk wisdom of the old biology on its head. Their yields are enhanced and their growth cycles are shortened, because (or although) they thrive in arid regions and on saline soil, are self-fertilizing, and draw the nitrogen they need directly from the air while automatically filtering out other harmful substances. They constitute a kind of objective test of conscience, for they are poisonous to vermin, but extremely nutritious for humans.

Animals are transformed into gigantic meat factories, provisionally still on four legs. The giant pig, infused with a few human growth hormones, is already alive and kicking on the drawing board. Thus the sausage we consume will be the first step to emancipation from the taboo against cannibalism, under which mankind has laboured for millennia.

The environmental problems that still threaten us on all fronts will see the advance of genetically manipulated microbes, able literally to gobble up harmful and toxic substances of all kinds, and will raise the white flag. One can only say how fortunate it is that we are so well provided with diverse poisons; otherwise we would not know what to do with the microbes that spring fully armed from the brains of genetic engineers.

Was environmental contamination and its discovery merely the anticipatory marketing ploy of an advertising agency with the coming biotechnological paradise on its books? Or are the slightly too assured declarations of victory only intended to drown out the background noise of terror, gripping even promoters and investors in spite of their rage to construct and their investments of millions, so that the choruses of approval also represent a way of keeping one's spirits up – by whistling in the dark forests of uncertainty?

On one side, the application of knowledge gleaned from animal genetics to human beings is coming up against ethical and legal barriers. In Germany as elsewhere, human cloning and the creation of man-beast hybrids are criminal offences. On the other side stand the worldwide pioneers of human-genetic reproduction engineering: agronomists, veterinary specialists and physiologists. Jacques Testart, fêted as the 'father' of the first French test-tube baby, has become an outspoken critic of this development. He records: 'The Englishman, Robert Edwards, was previously working on mice, and the Australian, Alan Trouson, worked on sheep and cows; I worked on cows and pigs. We developed these techniques and the doctors then turned to us, because they wanted to profit from what we had achieved with cows and other animals.' Testart sees a coercive chain emerging.

If ovulation is stimulated by means of hormones, a large number of eggs will be produced simultaneously. If the natural cycle is replaced in this way by an artificial one, freezing techniques must be developed for the embryo. Egg donations thus become feasible: women donate their eggs to other women. From my experience with cows I know what will come next: determining the sex of the embryo. With cows that is an economic consideration. But I can imagine couples asking me for a boy or a girl. The more people profit from my researches, the more this development worries me. (1988, pp. 65ff)

The grey area between animal and 'human' experiments is correspondingly large (Wollschläger 1987). For the fertilized, fissiparous cell group that will grow into a little person, subject to the permission of adults and the will of nature, is not legally a person or a thing. Theoretically speaking, an embryo is nobody's property. There is a joke at certain clinics, and perhaps it is not all that unrealistic: an embryo whose parents met only in the test-tube belongs to the doctor.

Take the German case as a typical example. Leading scientists consider research into living embryos, within fourteen days of the fusion of egg and sperm, justifiable. This position is supported by the German Research Council. Yet the public was able to learn that, according to expert estimates, 'at least 200 experiments on live human embryos at an early stage of development were required to create the first test-tube baby, Louise Brown', and that 'for three or four years now, eight- to twelve-day-old embryos have been tested to destruction.' 'Utility embryos' is the unintentionally harsh official coinage for these beings, upon whom research is allowed to deploy its techniques.

Such experiments are defensible, an expert committee argued, provided that 'they aid the detection, prevention or correction of disease in the relevant embryo, or enable the acquisition of well-defined, high-grade medical knowledge' – in other words, always.[1]

A central committee on ethics together with some regional committees, set up by the German Medical Association, is supposed to decide on appropriate research proposals. In reply to a question, an executive member of this committee on ethics said that permission was granted, in one case brought in 1986, for the egg cells of hamsters to be brought into contact with human semen. The emergence of a human-hamster hybrid was ruled out as impossible. The official moral philosopher added, clearly without realizing what he was saying: 'Many similar experiments carried out worldwide have confirmed this' (*Süddeutsche Zeitung*, 10 November 1987).

Admittedly, the application to the human genetic code of what has already been practised on microbes is still a gleam in the eye of researchers. Yet the prenatal genetic test marks the first step towards human genetic engineering. Molecular biologists can diagnose hereditary diseases such as muscoviscidosis, haemophilia, Huntingdon's syndrome (St Vitus' dance) and muscular atrophy, and 'prevent' them in combination with legalized abortion. It goes without saying that doctors are able to determine today whether a foetus is male or female, and whether it displays any chromosome anomalies or hereditary diseases. Soon, it is hoped, predispositions to schizophrenia or even criminality will be testable. So far 'only' the 'tentative pregnancy' (Rothman 1986) has been slipped past the German parliament and into law in this way. That means that the parents are not only entitled but obliged to decide whether they want to allow the child to be born if, from the point of view of genetic engineering, it is 'flawed' in this or that respect. Children are thus no longer simply conceived and born, but diagnosed for their qualities before birth, and perhaps eliminated. Parenthood is extended by genetic (i.e. 'divine technological') means, and supplemented by a bioconstructional element – initially by negative selection combined with legalized abortion.

As Elisabeth Beck-Gernsheim has observed, 'Like any other technology, [reproductive medicine creates] its own market.' A typical pattern is beginning to emerge: new biomedical aids are first introduced in order to stop or relieve suffering for a closely defined range of undisputed 'problems'. Then the phase of transition and habituation sets in, during which the field of application is continually extended. The end is foreseeable: all men and women are defined as potential clients – no longer, of course, to prevent direct damage to health, but because of the 'greater effectiveness' of technological intervention over nature's whims, incalculability and susceptibility to interference.

Thus the circumstances where in-vitro fertilization is indicated have multiplied within a few years, and are now so blurred as to be virtually unlimited. In-vitro fertilization with subsequent evaluation of the embryo has already been presented as the ideal method of the future to prevent severe disabilities at birth – or, more precisely, to prevent severely disabled children from being born. The use of inferior human sperm also becomes problematic. Only one man in ten would pass the quality controls applied in beef production (and bulls with inferior sperm are sent to the abattoir). In-vitro fertilization and embryo refrigeration are already being proposed as methods for timetabling the intervals between births. Increasing numbers of men are depositing their sperm at the sperm bank before having a vasectomy, 'just in case'. Married couples are already asking for another man to provide sperm, as they are dissatisfied with the appearance or personality of the husband. Similarly wives have asked for the eggs of other women, because they were not satisfied with themselves in one way or another. Reproduction technologies in general are being lauded as the royal road to family planning: 'Parents may soon be in a position to carry out total family planning, from control of family size to the sexes of the children and the sequence of males and females' (Beck-Gernsheim 1987b, pp. 287ff).

Prenatal diagnostics transforms the process of being born into an embryonic obstacle race. Every 'advance' increases the number of obstacles. This involves the intervention not only of parental choice at the previously untouched natural foundations of human life, but also that of social selection and thus of principles, ends, interests and prejudices. All of this takes place in a realm before the beginning of human existence, in the twilight of becoming where the limits of the medical definition of human life are constantly redrawn.

This development is still in the diagnostic stage. Yet some of our genetic do-gooders have intimations of higher things. One imagines that the 'gene repair shop' will soon be open for business: surgery will be carried out on the chromosome nucleus in order to replace 'diseased' genes by 'healthy' ones. As yet, the intention is corrective rather than creative. But that (potentially) leads down the slippery slope to changing a given DNA pattern by the addition, subtraction and permutation of elements. In this way the genetic text is, as it were, rewritten. This is in principle possible by means of a complete reading (or even an incomplete one) in conjunction with appropriate microtechniques. In the final analysis, it is a kind of genetic architecture (Hohlfeld 1988). This leads to new types, artful variations and whole new species of living beings. As said above, these techniques are already being applied with success to microbes, and we have them to thank for the 'green revolution' that is now inescapable. It is true that there are taboos that bar their application to humans. However, their scope will be greater than that, for instance, of all the family laws that parliaments have debated and voted on for decades. But there is a large grey area, and the boundaries will be defined by medicine.

Giving progress the benefit of the doubt: the inequality of the burdens of proof

Does not the subjection of human nature to human and social ends open up unimagined possibilities of self-creation for mankind and society? Have the biologists not given us the keys to a genetic reform of humanity that will make all previous political reforms and revolutions look like desultory scratching on the surface? Are we not in transition from stone-age politics to the genetic-preventive biopolicies of the future, policies which will further social ends through a change, as yet unimagined, of our central biological rather than our social structures?

An examination of the present state of research, with its protestations of innocence in these matters, will not convince many. It all adds up to an inequality: one side may continue its researches, no matter how far beyond the pale, while the other may not even ask any further questions. And is not the crucial point, this time at least, that questions should make clear the possibilities before these become actualities and thereby render questioning pointless?

The matter is substantially decided in advance by the choice of questions discussed. Those who consider the social consequences of human genetic engineering to be unknowable until they have actually occurred have to wait before they can act, or even think about it, and are thus obliged to acquiesce. Anyone who concentrates on the details of possible repercussions in isolated areas of policy entirely loses sight of the larger issue of the cultural consequences: 'Life, hitherto surrounded at least by traces of an almost religious aura of inviolability, becomes as technologically manipulable as plastic. Is that appropriate?' (van den Daele 1987, p. 42).

Whereas questions of detail can be solved at least in outline by the usual methods of specialist science, this gigantic question forces one to reflect upon culture and juggle with foundations and assumptions, and in turn perhaps diminishes the political relevance of recommendations. But even if one decides in favour of a microanalysis of the consequences, one has already answered the long-term questions, without having raised them, 'implicitly in terms of the currently established valuations' (van den Daele 1987, p. 42).

What is now fermenting in the genetic research laboratories is undoubtedly putting the 'essence' of humans, their very humanity, within the scope of social purposes and structural principles. 'It would be a metaphysical break with the normative "essence" of humankind, and, considering the complete unpredictability of the consequences, the most foolhardy game of chance – a blind and presumptuous demiurge hacking about at the heart of creation' (Jonas 1985, p. 197). Yet will the commitment to 'technological abstinence', however well founded ethically, be able to halt curiosity, ambition and the eventual deproblematization of research and medico-economic interests? Is one not permitted, indeed obliged, at least to know what one is up against – in other words, to cross the boundary before saying a qualified yes or no?

In the debate that has unfolded and will surely ignite over the next few years, uncertainty and ignorance about the consequences prevail on every side. However, while all the parties flail about blindly in the mists of the future, the burdens of proof and the opportunities for action are extremely unequally distributed. The geneticists and physicians can pursue a policy of the *fait accompli*, and are always well beyond whatever is still referred to, in public and by commissions of inquiry, as the very latest development. Furthermore, they are partisan – for genetic engineering must tailor its researches to the requirements of subsequent applications. Freedom to carry out research presupposes the freedom to apply it. At the very least, that means the unhindered idolization of research technique and development. Here, for the sake of pure curiosity, assent must be elicited for the application of research – not only before one knows what is being done and what the repercussions are, but as a precondition for posing the question in the first place. Hence the compulsive optimism, the preliminary skirmish on the subject of embryo research, and the many crude safeguards intended to secure the freedom to apply the results of research. Necessity is the mother of all these misbegotten measures, the need to minimize the consequences, suppress them, and make them invisible. It is the precondition for the very possibility of research. I say this in order to encourage a receptive mood for the strategic naivety on offer here.

The argument runs as follows: first, we will solve the central problems of humankind (famine, environmental despoliation, scarcity of energy resources, inherited diseases, Aids) by unprecedented means; second, these are really traditional methods, no different in principle from those of cheese manufacture through the centuries; third, nothing can go wrong anyway because we all have the very best intentions. All talk of monsters or eugenics is ludicrous and only stirs up the irrational fears of the populace.

Now this is a catalogue of fairly wild claims. Yet suppose we accept the promises at face value. Let us leave aside the fact that they are just as unexamined as the critics' fears. Let us forget that they derive from the poverty of an argument that says 'application' where it means 'research'. One might as well justify the construction of nuclear reactors by talking about the manufacture of pliers, or sanction capitalism by pointing to the commerce of the ancient Greeks.

It may well be difficult to locate the dividing line between such things as artificial limbs, false teeth, pacemakers, kidney transplants and genetic engineering; or between in-vitro fertilization, prenatal diagnosis and the recombination through genetic engineering of unrelated genera at the molecular level. (Not least because the disjunction is perhaps not discernible in the laboratory, but only at the level of social consequences; the latter, however, are circumspectly left unaddressed.) It is also the case that this perspective distracts one from the decisive question of what reality the test-tube holds in store. The future is wrapped up in the past in order to be brought – in the wrapping paper of the apparently familiar – unseen and unlegitimated, over the thresholds of public sensitivity. The Customs call that sort of thing smuggling. We all smuggle now and again. But to smuggle the genetic age beyond the limits of the supportable – in this late scientific era, such an exploit would be more than worthy of the Scarlet Pimpernel.

One may well take seriously this recourse to a display of 'good intentions' and the 'free will' of all the parties involved – and therefore one need not take it seriously at all. The road to hell, as is well known, is paved with good intentions: the 'good intentions' of the players are not enough to legitimate a technological chasm opening onto a veritable kingdom of creation and commercial transformation. This idealism is really not acceptable after 200 years of industrialism. It is equivalent to the conviction, positively wicked in its naivety, that the course of technological development and of its social ramifications was determined by the intentions of the founding fathers of science or by the voluntary involvement of its clients. In fact these proclamations are in blatant contradiction to the manifest autonomization of development, which can be gauged from such parameters as investment levels, international research competition, current and planned systems of research co-operation, the pre-emptive development of markets, etc. Under these conditions, public attention to the question whether it is right, permissible or what we want, is equivalent to Saint-Exupéry's Little Prince considering whether he should let the sun rise.

Thus the prudishness of the genetic engineers has strategic undertones. On this occasion, downplaying the issue is good for business. Moreover, here is a typical case of technological double standards: only the representations are chaste, not the deeds themselves. It may, therefore, be time to stir up the dust of possibilities on the horizon of the present.

Social consequences: will genetic techniques replace social policies?

'Good will' and 'free will' are formal concepts that can be given very diverse kinds of content. One would expect institutional problems, hitherto considerably neglected and underestimated, to play a role in this process. They are as it were the social hunger which, in combination with the scent of possibilities exuded by human genetics, stimulates an appetite for it.

The decisive potential of reproductive medicine and human genetics for social theory and social history consists in the replacement of social 'solutions' by genetic techniques. If access to it is granted, the field of operations of the genetic text quietly promises to provide value for money, encompassing and thoroughgoing preventive 'solutions' to social problems and conflicts which cannot be cured, but only massaged locally, by established political means. The dream of being able to 'plan' and 'realize' a life 'without problems' is nourished by technological possibilities (Rothschild 1989). That means genetic policies instead of social, educational, family and environmental policies – a real *deus ex machina*.

The following is perhaps not a completely unreal scenario: birth rates are in continuous decline in the western industrialized states. This is likely to have repercussions on every area of society: from the domestic market and the population's age structure and pension security right down to the military deterrent. Social-scientific research shows that parenthood is increasingly turning into a 'field of conflict' between divergent interests: on the one hand, the needs of married couples to further their careers and maintain mobility; and, on the other, the obligations and ties of parenthood, which can no longer be loaded onto an unprotesting wife. Would it not be an obvious measure, one that benefited everyone in addition to its desirability for women breaking out of past roles, to detach parenthood from the traditional 'genetic lottery' of love and partnership, and to arrange it separately – or, at least, to allow freedom of choice? The physicians have already managed to consummate marriages in vitro. Birth, too, has long been taking place in hospitals. Why not also shift the intermediate phase of gestation – above all, if so much depends on it: the emancipation of women and the economic, cultural and military welfare of all? Where is the difference in principle from what is already happening now, and how can one justify the claim that the old ways are better or more valuable than that 'extra bit of democracy' which we can take a genetic chance on now?

Even if this still comes up against 'problems of acceptance' today, is this resistance not ultimately just as irrational as the earlier reluctance, now overcome, to accept artificial hearts and kidney transplants? And how will this resistance be assessed socially if our populations, like the traditional pattern of love, marriage and parenthood, really do become extinct? Would it not then be thought a sin against our cultural inheritance that we failed to exploit the new clinical possibilities?

Perhaps the geneticists will succeed in raising, so to speak genetically, the maximum permissible levels of nuclear contamination, and also in changing our cell nuclei, so that atomic radiation of higher intensity causes itching: in that way, we will once again be able to recognize when danger threatens. Would it not then make sense to combine the two options, and to enact the healthy compromise of a 'basket of measures' in order to improve living conditions in the atomic age?

Is it not urgently necessary to grant researchers the right to experiment on human embryos (of course under the strictest conditions and the strictest controls of professional ethics), since animal experiments have no predictive power in relation to human beings? For, the argument runs, only thus can early, precautionary tests be carried out on the effects of certain substances on human beings, tests which are in themselves a precondition for the scientific legitimation of limits? Does not the opportunity present itself – from a historical perspective – of killing at least three birds with one stone: providing cheap, compatible and permanently available second organs for every man and woman, which would improve health and life expectancy; greatly increasing medicine's capacity to fight disease, and its diagnostic validity; and reducing animal experiments, quite rightly subject to vociferous protest in recent years – perhaps abolishing them entirely by switching to human embryo research?

Techniques for the early detection of hereditary and supposedly hereditary diseases are daily added to the armoury of prenatal diagnosis. Would it not constitute something of a grand coalition – between unborn children, who would be otherwise be saddled with these defects and the social conditions to which they give rise; parents, who would have to bear the additional burden; and insurance companies, which would have to foot (part of) the bill – to prevent babies 'unworthy of life' in good time, to winnow them out? Perhaps putting them to good use as organ donors?

In all modern societies there are jobs that require particular, exceptional abilities of one kind or another. Thus, for example, air safety depends on the fastest possible reaction times of pilots, and working with toxins requires specialized resistance to these substances. What objection could be raised against the genetic provision of these characteristics which would, after all, benefit everyone? Let us assume that allergies will continue to increase with the universal dissemination of harmful and toxic substances in air, water and foodstuffs. Would it not be an obvious step, and one precisely in the interests of coming generations, to fight by genetic means, as a kind of hereditary disease, the human predisposition to allergies – not least because we can thus save ourselves the trouble of closing the floodgates of poison by political means?

Medicine is constantly forging ahead. There are successful transplants of one organ after another; there is only a certain shortage of suitable organs. Hitherto, even children from the Third World have sometimes had to earn another crust by selling a kidney to the more affluent people of first-world countries. Might one not deal with this intolerable situation, while considerably cutting expenditure, by granting everyone the right to cultivate and deep-freeze replacement organs, as this hurts no one and only marginally extends the scope of today's sperm and ovum banks?

Some are bound to cry out that this is 'barbarism modernized'. But what are their arguments? Soberly appraised, they represent a particular value system that they do not even disclose. They ought first to display its credentials and defend it against the various possible criticisms. Only a subtly differentiated, balanced attitude will help here. Rhetorical terms like 'barbarism' are inappropriate, if only because they deny moral integrity and goodwill to those (and God knows there are enough of them) who propose the measures. Think of the infirm, and of the diseases that can be got rid of in this way . . .

Besides, the imagination has only been invoked one-sidedly, restrictedly. Has it not merely blown up the evils into a bogey? Why should cloning, for example, be without its charms? Perhaps it is unattractive to biologists, but not to twin researchers, ethnologists, sociologists and psychologists. Does this not open up a new field of social and scientific experiment on human identity, the question of 'who am I?' Does not biotechnology provide the first-ever opportunity of detaching the problem of identity from the natural substance that embodies it, and thus of being able independently to vary it and detach it from the biological background upon which it apparently supervenes? Does it not also enable philosophy to start afresh after its long, moribund years as an exhibit in the museum of its own history? A philosophy that might be accessible both to empirical fact and to social experience? 'In other words, would a society that permitted and even favoured the existence of several clones be more barbaric or less free for that reason alone? Might the regression not rather consist in preserving a state of society where the possibility of individuation and of socialization is dependent on the uniqueness of the uniquely produced natural organism?' (Koch 1987, p. 72).

Molecular biology has opened up a new area of direct social policy. This makes possible a change in the constitution of life through genetic technology, not subject to law or to problems of enactment, interpretation and implementation that happen to be the norm in democratic societies; a constitutional change to the subject–object relations of human life, thus rendering incapable of resistance those persons changed, preventively and irreversibly, to the marrow. Irrespective of the assurances of today's politicians and the wide-eyed innocence of researchers, all this constitutes the decisive political temptation. The latter grows as the problems get out of hand, policies become powerless and universal uncertainty about values is produced by a modernity that both deconsecrates and is deconsecrated. If, for reasons I cannot at present conceive, we were to witness a new Christian miracle of abstinence in the genetic field, the flood gates would first open elsewhere, and then here. For the right to ignorance has been lost once and for all. In the face of technological temptation, chastity would not only have to be perpetually asserted, but preserved – under conditions of global competition and in the institutionalized mistrust that binds nations together.

The human-genetic 'Dialectic of Enlightenment'

Horkheimer and Adorno saw a certain inner inevitability in the transformation of the 'Dialectic of Enlightenment' into the racial madness of German fascism. There was always something artificial about their argument, not only because the antithesis between enlightenment and concentration camps seemed too huge, the change too radically conceived, or because explanation after the event loses much of its power to convince. Above all, however, the guilt thus appears to be apportioned far too generally. The effect is almost that of a false absolution if the Enlightenment, reduced to technology in bourgeois society, suddenly changes into its 'logical opposite', in German concentration camps of all places. Why, for example, did

Germans make no response to their countrymen's brutalities, while even the conquered Danes mustered some resistance in acts of collective civil disobedience against the barbaric outrages? Perhaps today this falsehood as to particulars can more easily be separated from the general correctness of the theory. Perhaps, after all, the reasons for the Nazi crimes perpetrated against Jews and others, but also covered up in their cool, bureaucratic implementation of prevailing ideologies, should first be sought in the seamy side of German history. All the same, Adorno and Horkheimer's line of argument makes it clear that human genetic engineering is not owed to some historical accident or incident, but must be understood to issue from the Enlightenment as applied to technology. The project of the technological subjugation and perfection of nature, thought through to its end and realized, must sooner or later encompass human nature (and 'later' means 'now'). Mastery of nature and of the subject thereby overlap and reinforce one another in a specific sense. First, as always happens when technology opens up a breach, nature can be chosen, shaped. Also, the privilege of not making a decision is abolished, as is the option of using what is to hand in the search for ends. Human nature is now subjugated to ends that are bound to issue from the hearts, brains and political ideologies of human beings, even if this daring advance in the shaping of nature is carried out under the best-legitimated banners, for example that of preventive health care.

Yet what used to be valid is no longer sufficient for a grasp of the simultaneous mastery over nature and subject made possible by human genetics. Here human existence is placed at the disposal of society and technology; prenatally, to be sure, and to that extent 'not yet in existence', but with its subjective modes of expression already present and at society's disposal. Human genetics has to do with substrates that enable access to subjectivity. Common talk of fighting hereditary diseases is more than a linguistic error. In fact, the sufferers from hereditary diseases will be done away with in this way – people, in other words, who certainly do not want to see their lives and experiences reduced to this sole characteristic. This only apparently subtle distinction must be exposed as the very opposite . This must be done not only because the label of 'hereditary disease' will come to represent a death sentence in the human-genetic workshops of the future, and because there is here an extreme form of reification, where human existence is identified with a single disease symptom, but also because the difference between these two concepts contains the worlds that separate the history of surgery up to the present day from the creationist surgery of the genetic age.

The classical surgeon cuts and snips legs, hearts, kidneys and so forth, but is always dealing with the living, anaesthetized human body. Genetic engineers, by contrast, deal only with substances, test-tubes and formulae, yet in spite of this level of abstraction they penetrate human subjectivity directly. Mastery over nature coincides with mastery over the subject, the two being short-circuited and rendered operable by (and as) matter. Genetics is a technology of the future that shapes the material substrate of the modes of expression of future living things. Whoever can thus change whole generations is no longer dealing with a physically embodied person, but with apparently dead matter that can be selected and instrumentalized at will, 'without pain', according to determinate disease symptoms.[2] To give an example: a researcher trying to overcome schizophrenia by genetic means will, if successful, have carried out precisely the stated ends of the eugenic movement of the nineteenth and early twentieth centuries. These were realizable at the time, if at all, only by unbelievable cruelty: by getting rid of sufferers from schizophrenia. The effect is the same, indeed it is incomparably more powerful and efficient, while the path taken is an entirely different one, as for the most part are the guiding ideas.

This human-genetic control of the subject, naturally mediated, elevates the techniques practised here to a new significance in the history of society. Human modes of life become accessible to technology, changeable, through the labelling of diseases – not in their present state, but potentially. To talk of 'hereditary genetic disease' is objectively to further the cause of eugenics, though by incomparably more elegant and effective means. The dialectic of the Enlightenment thereby gains a new meaning for the future: human genetics is in its basic conception a mode of knowledge that aims at, indeed compels, eugenic praxis.

To be sure, no political manifesto advocates Huxley's *Brave New World*. There is public outcry against the nightmare visions of 'advances in genetic engineering' as depicted by Nobel prizewinners at the now notorious London conference (cf. Löw 1985, p. 186ff and others). Almost everyone, surely, deploys, researches and promulgates the new techniques in the firm belief that the racial theories and their eugenic madness have disappeared for ever in the mass graves of the Nazi terror. But eugenic theories of race and their political executors only disposed of primitive techniques – compulsory sterilization, laws against mixed marriages, and mass killings – which had to be applied to living human beings through hate propaganda, social hysteria and brutal, giant bureaucracies (from legislative procedures to ordinances for crematoria); the new genetic technologies are eugenic by virtue of their operational logic. For the moment, this much is true: there will be no eugenic policy, transforming a country into a band of murderers that hounds and exterminates people of other races and creeds, and into those who countenance such murder – not, in any case, through the achievements of genetic engineering. There will be no racial theory, nor any eugenic movement (such as was very influential at the beginning of this century, e.g. on individual state legislation in the USA), if only because the new technologies facilitate the practice of eugenics without the use of force. It is a practice grounded in the technological change, now freed of ideology and repressive bureaucracy, clinically neutral and realizable in vitro, of the genetic equipment of life in all its forms.

According to the still prevalent ideology, technology is neutral: its application alone determines whether it serves 'good' or 'evil' ends. If this self-protective claim of science, this defensive belief in the unspotted reputation of technical achievements, lost some of its plausibility when the atom bomb was dropped, genetic engineering flatly refutes it. Every encroachment on the inherited substance of plant, animal and human life by successes in genetic research means a practical selection, and thus the end of the line for certain developmental variants, and the breeding or improvement of one or other aspect (health, performance, etc.) of other variants. In other words, statements concerning the structure of cell nuclei bear within themselves their own application, their aim being always, independently of concrete purposes, contexts and intentions, the selection, breeding, improvement, recombination (in other words the shaping or creation) of living beings, including humans.

Human genetics is a mode of self-encounter between humanity and the project of its history. During the course of science, humanity has cast itself as mechanism, and now it is discovering at its core a formula, a blend of chemical substances and biological cell structures. In the process, no actual human being has come to the surface or leapt out. Therefore, it is argued, humanity cannot come to any harm.

The abstraction of the laboratory entails commerce with the banality of chemical substances. These stand in the same abstract relation to the thought of a living thing as the formulae that express them. For this reason, boundaries between life and death may be shifted arbitrarily, nominally, without reference to experience. Nothing suffers pain, nothing responds, nothing defends itself. All talk of eugenics is accordingly the meaningless drivel of stone-age fanatics who do not yet realize that the kind of human for whom they are struggling exists only as a legend, a refuted working hypothesis. Life? A pious phrase, a necessary public relations exercise, the more freely to pursue, behind the raised sightscreens of responsibility, research being carried out anyway. It is precisely this impalpability that predestines the unity of research and practice in genetic engineering to develop into the concentration camps of the future, even if these are only comparable in their function to those already known. Here nobody can know any longer what he is doing. So everything will run smoothly. It is no longer necessary to forget and to repress, as knowledge was never at hand in the first place.

The eugenics of the future will not be applied directly to body and soul by means of laws and gas. It will have divested itself of this horrific early form of a total social eugenics. The abstract test-tube eugenics of the future will avoid discrimination and killing by preventive technical means. Its representatives need know, believe and say nothing of racial theories, for technology itself silently and effectively carries out the 'amelioration' of life without the need for ideological and social hysteria. Thus we no longer have to deal with a political or social eugenics, but 'only' with a technological eugenics which, neutrally enforced, keeps quiet about its purposes. For precisely this reason, by the criteria and principles of industrial mass production, it can celebrate its victory today, in the reassuring guise of preventive health care and with the scientific blessing of genetic counselling.

Barbarism may supervene because it does not appear on the political stage, clothed in the familiar garb of brutality. It gains access through the clinics, laboratories and factories of the new biochemical industries. Its victory parade does not begin with street brawls, the persecution of minorities or people's assemblies, the dissolution of parliaments and abolition of constitutions. This time it steps onto the stage of world history dressed in white coats, of self-confident research, the good intentions of doctors and the desire of parents to do their 'best' for their children. Lawcourts and parliaments assent because everyone bursts into applause when the banners of health are waved aloft. Who would want to impede the genetic struggle against hereditary disease and thus prolong suffering? What is there so bad about parents wanting to have 'healthy' babies? What terrors can efforts to create higher-yield plant and animal species hold? The eugenics that threatens has shed all the distinguishing marks of a sinister conspiracy, and donned the robes of health, productivity, the promise of profit. It is given impetus by the dramatic hazards it is supposed to address (Aids, ecological destruction) and by the power of investments worth billions in 'commercial eugenics' (Rifkin 1985).

The dream-child mentality:
the gentle compulsion to have a perfect child

Free will prevails in western democracies (we shall not deal here with what threatens in the wake of genetic engineering's victory in countries and cultural environments with different political systems, values, ideologies). Yet the phrase 'free will' has the old, scornful ring. It connotes a lost social consciousness, a forgotten, ineffectual sociological enlightenment, as if there were no institutions or interests capable of pressing 'free will' into the service of their own ends, or as if 'free will' were a way out for the distress of parents, who later either justify to a child his or her 'hereditary disease' or 'disablement' (as one used to justify oneself before God in silent prayer), or become, by virtue of their assent, the executors of a (new) eugenics which will often enough discriminate against them as carriers of the same 'defect'. Once the state of the art determines the norm, abstention from choice becomes a luxury that no one possesses any longer. The dilemma of having to decide, and of not being able to decide, between yes and no unfolds with inexorable rigour. The helpless parents find themselves once more burdened, one way or another, with the unconscionable responsibility of the godlike role of creation that accrues and is assigned to them through technology.

The first social filter of 'free will' then also lies in what has rightly been called the 'dream-child mentality'. All parents want the best for their children. Their personal wishes manifest themselves in this generality as subjective metastases of social values whose causes are manifold, including the hindrances and allurements of the parent–child relationship in the disenchanted modern age. If the best used to be 'health' and 'education', from now on it is 'preventive genetics'.

'Responsible parents of the future', writes Elisabeth Beck-Gernsheim, 'will have to ask themselves whether their own "hereditary substance" is equal to the demands of the age, or whether they had not better use the expedient of donated sperm and ova – carefully selected, of course' (1988b, pp. 105ff). The moral philosopher Reinhard Löw paints a provocative vision of the future: 'In this brave new world, having one's own children means sending them on their way with the unconscionable disadvantage of lower intelligence and a less attractive appearance than those produced by progressive means, or combined in a test-tube. One can almost envisage the time when children will complain to their parents of their "defective genetic inheritance"' (1985, p. 179).

Future embryonic and pre-embryonic 'quality control' practices are already observable today in the selection of prospective sperm donors and surrogate mothers.

> An artificial insemination specialist who practises in Essen gives the follow-ing criteria, among others, for the selection of sperm donors: 'No pro-truding ears or hooked noses', minimum height 1.75 m, no 'weirdos'; and 'stable family backgrounds'. Such procedures are . . . not in any sense intended to be eugenic, or to represent selective breeding . . . but what does 'unacceptable' to the 'clients' actually mean here? A black sperm donor for whites, for example? What is a defect? Protruding ears, for example? . . . and whatever and whoever counts as 'bad' in this sense also substantially depends on social yardsticks and evaluations. (Beck-Gernsheim 1988b, p. 107)

To parents, the development of human-genetic reparability and creativity means that their aims and desires, in line with social norms, can now be realized more efficiently; even, at the pre-embryonic stage, preventively. The parental concerns of the past about choice of baby food, breastfeeding times and parental help with children's home-work may, if everything continues as it is, be rounded off in the future by genetic screening, genetic counselling and the operational selection of the 'career gene', in order to decide the children's future in the competitive social struggle. The choice of education and occupation would then be completed, circumvented and reinforced by 'natural parental choice'.

In this way, ideas of 'a life worth living' and 'a life not worth living', still dormant in the collective subconscious, can be awakened by the scent of technological possibilities . Human prejudices against humans can now avail themselves of the final, genetic-surgical solution, under the apparently neutral label of 'hereditary biological defects', not in order to 'conquer' the victims of their unmastered fears – that would be to speak too harshly, and also too weakly – but in order to prevent their even being born, on the path to the perfec-tion of the human race. This danger of a medically implicit eugenics grows with the scope of medical diagnosis and surgery, and in the grey areas and social valuations which lurk there. With infuriating insensitivity politicians have already recommended 'preventing the inheritance of grave hereditary illnesses by ascertaining the family history of the sperm donor' (Bundesminister der Justiz 1987, p. 34).

In addition, the cause of health is given carte blanche by 'free will', virtually excluding an unfavourable decision. It is almost certain to have its way, in view of the social costs that accrue from the 'luxury' of supporting the victims of hereditary disease. Health lubricates 'free will', making it comply with 'necessity'. Thus the health question turns into a prenatal death sentence, straightening out the 'crooked timber', as Kant once called humans, in the mild, silent, clinically sterile manner.

Talk of free will thus misconceives the relationship between social values or institutions and new technologies, truncating them into a private relation. The most powerless is supposed to be omnipotent. In the process, however, the paradigm issues that might disturb and arrest the onward march of human genetics are foisted on the individual (in)capacity to decide. Yet the offical myth of 'free will' is exploded, at the latest when the social structures thus introduced and honed are displaced. Always assuming that we do not change overnight into a society of angels, living only by the precepts of pure, brotherly love – that is, assuming that what has been true till now will remain true tomorrow, there are very good grounds for assuming that human-genetic selection will intervene – never mind free will – wherever costs can be cut.

If costs are tied to parents' prior genetic provision for their children's intelligence, then 'education' and 'performance' will be sociobiologically formatted. Social inequality will be established anew and biologically 'justified' in terms of parents' disposable income, through the medium of socially manufacturable 'high-performance' genes. And the insurance companies will stipulate the costs that are to be borne by the policy-holder if 'genetic reason' does not prevail.

Who will prevent the entrepreneurs and their personnel departments from supplementing their aptitude tests with genetic tests, if they can avoid lost working time, sickness benefit payments and friction within the organization? Perceptions of human gene 'defects' harden into discriminatory labour practices, and in this way the pressure to choose the preventive genetic solution intensifies, though by now that is probably unnecessary. 'Genetic screening and gene-pooling will create a new variant of social inequalities and prejudices, not dissimilar to racial or sexual discrimination' (Rifkin 1985). These will be social barriers, emerging with and from genetic knowledge and its operational possibilities. If genetic screening and gene-pooling become widespread they may provide the foundations for a sociobiological class system with new, subtle distinctions between status hallmarks such as 'preventable hereditary' and '(as yet) unpreventable hereditary' defects.

2

The Naturalistic
Misunderstanding of the
Green Movement:
Environmental Critique
as Social Critique

The question of the desirability of 'advances in human genetics' hinges on social acceptance, rather than on medicine, nature, risks or repercussions. The decision must always be made, ultimately, on political or cultural grounds, and the very possibility of making the decision must indeed first be wrested from the preconceptions of technology. But such proofs do not constitute 'natural' arguments for one line of development or another. The sovereignty with which technical decisions are made and justified on other than technological grounds is one measure of democracy.

Nature: social memory as utopia

When anyone says the word 'nature', we should ask the question, 'Which nature?' Naturally fertilized cabbage? Nature as it is, industrially lacerated? Country life during the 1950s (as it is represented in retrospect today, or as it was represented in days gone by to countryfolk, or to those who dreamed of country life)? Mountain solitude before the publication of hikers' guides to deserted valleys? Nature as conceived by natural science? Nature without chemicals? The polished ecological models of interconnectedness? Nature as it is depicted in gardening manuals? Such nature as one yearns for (peace, a mountain stream, profound contemplation)? As it is praised and priced in the supermarkets of world solitude? Nature as a sight for sore eyes? The beauty of a Tuscan landscape – in other words, a highly cultivated art of nature? Or nature in the wild? The volcano before it erupts? The nature of early cultures, invested with demonic power, subjectivity and the living gods of religion? The primeval forest? Nature conceived as a zoo without cages? As it roars and rages in the cigarette advertisements of the city's cinemas?

Every human being is a part of nature too. Yet where does nature begin? When a babe is first suckled? Or as soon as a woman goes off the pill? In sexual intercourse (where and how)? Of the homosexual or heterosexual variety? Polygamously, perhaps, in selective diversity? Or in extramarital fidelity to permanent change?

Of course, the physique of the European adult on two legs is no longer that of nature pure and simple: beer, the nine-to-five office job, norms of dress (and undress), the clothing industries, job security in the cosmetics industry, and the idea that a human being must create himself in the image of others' expectations have left a few historical traces. Chest hairs sometimes recall the male to a natural past, and awaken dreams of natural possibilities which – in view of the bloatedness of the formerly natural human body, or the high muscular definition available to citizens in the gymnasia of our economic wonderland – make the cry of 'back to nature' thoroughly understandable.

The 'natural blend' with which we have to deal today, a remoulded nature devoid of nature, is the socially internalized furniture of the civilized world: work, production, government and science at once reconstruct it and furnish it with the norms by whose yardsticks it is judged to be endangered and damaged. The process of interaction with nature has consumed it, abolished it, and transformed it into a civilizing meta-reality that can no longer rid itself of the attributes of human (co-)creation. In this age of battery farms, plant, animal and human genetics, parks, development programmes, and the 'renaturalization' of towns, one is dealing with variants of an artificial nature: projections of nature, wish-fulfilment natures, nature utopias, all roughly as natural as a big-screen advertisement replete with roaring, turbulent rivers in the urban bustle of Tokyo.

This irreversible artificiality of nature is additionally, if unintentionally, confirmed precisely by its conservation through ecological intervention: thus there appear centrally administered museums of real nature, constructed according to ecological principles, 'arks of civilization' for dying natural species. In the mixed forest of administered nature that emerges here, dying species of songbirds and plants are offered an appropriate breeding ground for their civilization.

Another index of nature's transformation into society is the degree to which disasters, natural disasters of the 'classical' type – landslides, floods, dying forests, and so forth – are now interpreted and treated as disasters for which policy is answerable. That provides a measure not only of the social integration of nature, but also of social perception of this basic fact.

To put it paradoxically, the social 'consumption' of nature renders philosophically invalid all those concepts and theories that conceive of nature as the counter-image of human activity and power, to which it must be subjugated. At the same time, there is a vindication and revival of conceptions which model nature not as a world of dead objects (with God as the 'supreme mathematician'), but instead as living, intelligent and active, as in Schelling's philosophy of nature.

Thus even nature is not nature, but rather a concept, norm, memory, utopia, counter-image. Today more than ever, now that it no longer exists, 'nature' is being rediscovered and pampered. The ecology movement has fallen prey to a naturalistic misapprehension of itself: it reacts to a global fusion, rife with contradictions, of nature and society; this fusion has weakened these two concepts in a blend of reciprocal interconnections and injuries of which we have as yet not the faintest idea, let alone a proper conception. The high esteem in which it is held corresponds to the devastation and loss, and draws from memory a 'nature' that is anything but natural. This does not merely concern epistemological questions: opportunities for political action, arising from the despoliation of 'nature' as a domestic social phenomenon, are lost (cf. part II).

The very word 'nature' still seems to have a green flavour. The concept of nature is a self-negating human invention: with his internalized conception of nature, man abolishes and rescinds his own role of creator, discoverer, ruler, destroyer. More than that, man cultivates it in opposition to his role of creator and destroyer, upholding it as the extreme of non-alienation, of non-civilization. Recourse to the concept of nature gives the appearance of an outer limit, prescribed from within, to humanity's perceived subjection to increasing hazards and self-destructiveness. The concept of nature does not betray the model its formulator associates with it – to either the speaker or the listener – not, in any event, at first glance. It is, as it were, language that appears to retract its utterance, a concept that apparently 'leaps to the eye', really 'growing out' at the beholder. Its significance is that the concept of nature enables the speaker, by recourse to the external world, to set out what is inwardly, profoundly oppressive. The concept of nature achieves a kind of self-expropriation in which, as it were, the image becomes independent of the subject that it mirrors, and provides – in reflecting it back – the mirage of a reality for itself, apparently uninvolved in all the mirroring; a given, by whose means mirrored subjects can orientate themselves. The effect is internal, as are the model and the conditions for triggering it off. But the operation proceeds via an external world that pretends to an utterly self-evident self-identity.

'Nature' is a kind of anchor by whose means the ship of civilization, sailing over the open sea, conjures up, cultivates, its contrary: dry land, the harbour, the approaching reef. In the process it negotiates the conditions under which it may continue to voyage, to drift.

Upon close inspection, all who talk of 'nature' in the sense the word pretends to, namely that which is untouched, free of human creation and destruction, have always refuted themselves. To speak thus presupposes amnesia – of the fact that talk of nature conjures up the whole dichotomy, the history of nature's subjugation, cultivation and destruction, the history of concepts of nature – and it also begs the question of the sense in which the word 'nature' is used, when the subject under discussion is the shaping of life in society and the provision of social norms. The concept of nature simulates a naivety that allows its utterer to lay claim to a naivety of the given, of the prior given, the immutable and good, which becomes the more significant and enticing as doubt is cast on all unquestioned assumptions. 'Nature' seems capable, if not of cutting through the Gordian knot of civilization in which one feels bound up, entangled, then at least of helping one loosen and shake it off. It is, as it were, the bolt-hole of anti-modernism, keeping open to its dissidents (those weary of modernism and convinced anti-modernists alike) the option of modernism as a variant of itself.

The theme first secreted into the concept of an external 'nature', which, however, breaks forth ever more openly and directly, is that of self-delimitation: the self-guidance and self-determination of a modernity that has always yielded its claims of shaping reality to the flat necessity of a manufactured determinism of progress. There are manifold reminders of the fact that the meanings of 'nature' do not grow on trees, but must be constructed.

The history of the concepts of nature is the history of antithetical models and evaluations. Thus supporters and opponents alike of the French Revolution appealed to 'nature'. Robespierre activated the guillotine in the name of nature: 'the heads rolled, and the executioner called out after them: man is nature's noblest creature' (Lepenies 1989, p. 8). For Edmund Burke, on the contrary, 'everything seems out of nature in this strange chaos of levity and ferocity' that he felt the Revolution to be. 'We fear God, we look up with awe to kings; with affection to parliaments; with duty to magistrates ... Why? Because when such ideas are brought before our minds, it is *natural* to be so affected' (1953, pp. 58–9). Indeed the word had many significations and could be used by all political parties. Precisely therein lies its explosive force, the ideological power that emanates only from sufficiently diffuse concepts. By 1800, only too understandably after the experience of the Revolution, 'nature' had become perhaps the most dangerous word in the French language.

All scientific representations of nature as governed by laws are also projections. They may be true or false, but they are not of a nature that has, as it were, recovered itself in human consciousness. Nature does not speak to us even in experiments; rather, scientific questions are (more or less) answered. If the experiment says 'no' – thus apparently expressing nature's veto – then the interpretations and consequences of this 'reply' still remain to be decided entirely by the researcher.

Even ecology, the spokesperson for nature conceived as a network, is a variant of natural science, not nature's own articulation of itself. Moreover, it is a variant which was sickly for a long while (ever since Ernst Haeckel coined the term in 1868), and is now attempting to compensate for its century-long slumber by a kind of surprise attack which bears all the hallmarks of a cybernetic hyperscientism. To put it another way, ecology is on the verge of placing itself at the greatest possible distance from that 'naturalness' to which it sometimes appears disposed to give expression. Yet the attractiveness of ecology surely derives from an awareness of the repercussions of economically programmed, highly specialized, professional natural science and technology. Not least by averting their gaze from the connections that they destroy, these disciplines have become the motor of self-jeopardization of 'nature–society relationship' that is our concern. But the allure of ecology thereby answers precisely to a modern experience, which seeks for its articulation counter-images developed and jettisoned in the historical process – i.e. ecology. All the same, this is a variant of human conceptions of nature which, by thinking in relational concepts and norms, sensitizes one to the devastation to which the natural-scientific intoxication with technology blinds one.

Ecology is guilty of forgetting about society, just as social science and social theory are predicated on the forgetting of ecology. The terrain has been staked out by the concepts both of system and of environment. The proponents of each disdain the other, without noticing that it is public awareness that preselects ecological questions via a historical aggregate of society and nature, in which so-called 'ecological hazards' are always systemic hazards.

Murray Bookchin is speaking on behalf of many others when he writes (1982) that only ecology can provide the principal axis along which society should be organized in the future. For Bookchin, spontaneity in the social life of society converges with spontaneity in nature. What a happy coincidence that 'autonomy thereby becomes, in a sense, a law of nature (to the great delight of the ecologists)' (Lalonde 1978, p. 53). Nevertheless, the ecological crisis leads many to call for the 'strong hand' of the state, which, as Herbert Gruhl urges, should 'not merely conceive, but ruthlessly implement a strategy for survival' (1975, p. 305, quoted in Oechsle 1988, pp. 44ff; see also Harich 1975).

If these texts of the ecological *Sturm und Drang* movement of the 1970s deduce society by straightforward inversion from 'ecological laws', then influential concepts can easily be discovered now, which, in their sociological naivety, recommend 'ecology as a basis for action, for politics'.

> Since it is a scientific basis, capable of exact proof, testable, i.e. precisely adequate to the demands of occidental, rational, scientific method, no one should baulk at it . . . if ecology were only a subjective whim, if it did not derive from matter itself, if it were like a religion, how marvellously free we would be to decide! But ecology compels us. We cannot escape its insights with impunity. In this matter we are unfree, for it is a matter of laws of existence that extend beyond ourselves, and to which we are subject. (Maren-Griesbach 1982, p. 32; similarly Oswalt 1983; for a critique of this position see Trepl 1983 and 1987, and Oechsle 1988 for a summary)

On the other side, 'ecological blindness' is a congenital defect of sociologists. Sociology was only able to win and assert the independence of its discipline from the natural sciences by opposing a social discourse to that of science, or, if that is too crude, by concentrating upon the 'social facts' of Durkheim, the 'social praxis' of Max Weber, etc. (Luhmann 1989). Even if, for example, the Chicago School's project of 'socio-ecology' staked an early claim to the concept, this still cannot gloss over the fact that it is essentially concerned with specific investigations in empirical urban sociology, whose relationship to the environment remains rather marginal. Only in recent years has a series of social-scientific studies of the 'ecological crisis' been published, and these only haltingly acknowledge that access to ecological questions is certainly not to be gained without a basic theoretical change of attitude (cf. Bühl 1986; Wehling 1987 for a summary of 'ecological orientations in sociology').

It is intended here to illuminate in two stages the conversion of society into nature, the naturalization of social problems: first, the relationship between destruction and protest will be addressed; second, it is intended to pursue the question of why ecological protest appeared first and most markedly in West Germany (and less spectacularly in other developed industrial states).

The muteness of destruction and the sources of protest

Orgies of mathematics and science are held in defence of nature. Whole battalions of high-powered economic calculations advance, flanked on either side by dissidents from the natural sciences who wish to invert the formulae from which the hazards have escaped and make possible their recapture. This is indispensable, no doubt. Only thus can the institutionalized alarm systems be triggered off. Furthermore, the hazards which undermine our health as much as they seal the fate of endangered bird and plant species can often be brought to public notice only in this way. Only thus can the institutionalized concealment be confronted, on its own terms, with a little of its old truth. Yet it must be said that all these efforts are only a substitute, a strained way of saying, 'We do not want to live thus!'

The discussions of the last decade, once again displaying and depicting the whole arsenal of the critiques of technology and industry, have remained technocratic and naturalistic at their core. They exhaust themselves in exchanging and conjuring up pollutant levels in air, water and foodstuffs, and comparative statistics of population growth rates, energy consumption, food requirements, raw material shortages, etc. This is done so ardently and exclusively that one might think someone like Max Weber had never existed, and that he would appear to have wasted his time in demonstrating that these considerations are meaningless unless one takes into account the social structures of power and distribution, bureaucracies, prevalent norms and rationalities. An assumption has been smuggled in and established, placing modernity on the Procrustean bed of technology and nature (for which read culprit and victim). This inadvertently helps to bring about what it purports to oppose. Its approach prevents this way of thinking – which is shared by the political environmentalists – from grappling with the rules, the distribution of the burden of proof, and the power to act without certainty, or, in other words, the relations of definition which determine how big technological hazards are socially negotiated. Risks are predicated on cultural norms of acceptance, and also upon legal and scientific standards for assigning liability and compensation (see part II). Anyone who fails to appreciate this falls victim to a force-of-circumstance ideology, further shored up by anti-technocracy in technological clothing.

One of the hazards of hazards is the naturalization and technologization of social problems, which the social staging of hazards compels. As will be shown later, there are many advantages to be had from turning nature into the site, the theme of social discussion; but it is still a discussion that displaces its subjects and theatres, and for that reason not only continues to be unstable, but always threatens to skate off into the contrary of its intentions.

Ecological hazard indicators

Nothing, surely, illustrates so well the hazardous situation of a highly industrialized state as the significant fact that, in spite of more than twenty years of conscious environmental policies with a broad spectrum of activities and several successes, the overall destruction has at best been slowed down, though by no means halted, let alone reversed (A. Bechmann, 1987b).

Information agencies, advice centres, etc. have been set up one after another, mountains of laws have been passed, which now impose legal rules on virtually all environmental factors (excluding land protection). In the manufacturing industries, 4.6 per cent of investment is spent on environmental protection measures; these also account for 1.7 per cent of public expenditure (*Umwelt* 8, 1985). More than 200,000 people are employed in the environmental protection industry in Germany alone.

Yet all this has been unable to halt the destruction, which runs its course within the prevalent norms, and is thus broadly legitimated: the German government's environmental programme was never fulfilled (A. Bechmann, 1987a). There are any number of obstacles between the laws and their enforcement. The money that pours in is primarily compensatory rather than preventive in its effects. To put it figuratively, a new filtration plant is built while the source of the pollution is left intact. Sulphur dioxide emission levels – formerly thought to be an index of success – are still around 3 million tonnes per annum. Hydrocarbon emission levels, waste 'mountains' and damage to forests are all increasing (the latter is currently around 53 per cent, an unimaginably high increase on the figure for 1982).

This wealth of risks is not only given expression by the lists of endangered plant and animal species. It also undermines the health of the population.

Over the past thirty years, the percentage of deaths due to cancer has grown rapidly: from 15 to 23 per cent among men, from 17 to 25 per cent among women. Every tenth citizen of Germany is allergic to one or several foreign substances in the air, water or food; some 70 per cent of the population are already latently sensitized. Children in particular are increasingly susceptible to this. They are alarmingly more prone to contract cancer (Dost 1983, p. 166), skin diseases and asthma (SPD 1985, p. 2). To a high degree of statistical significance, all of these disease profiles point to the environment as catalyst, contributory factor or cause (Wassermann 1984, p. 52).

Germany, with its genius for exports, is in danger of choking on its own waste. Three hundred million tons of waste annually pour out of factories and homes, together with 5 million tons of toxic waste, much more than half of which is produced by the chemical industry. If current dumping practices continue (75 per cent of the toxic waste is dumped at twenty-four special sites), the available space will soon be filled up. New dumping sites are difficult to create, for political and technological reasons (Stiller 1987, p. 51). More and more rural communities are broken up, and roads laid down. Just like the chemical and nuclear plants where they occur, accidents are becoming an ever more certain bet. The new hyperbole of 'safest' is even safer, whence it follows that the only safe thing to say is that safety is impossible. According to a study by the Ministry of the Environment in West Germany, 1,000 non-nuclear chemical plants in West Germany are running a high risk of dysfunction (difficult to quantify by current control standards). Many of them are located in the middle of cities, so that, to judge from so-called 'chemical density', living in North-Rhine Westphalia is nine times as dangerous as living in the USA.

Many of the hazards are registered time and again, but continue to cause damage. At present some 8 million chemicals are registered in the Chemical Abstract Service, and around 300,000 are annually added to this list. Even if many of these substances are only short-lived under laboratory conditions, the number of synthetic substances is increasing inexorably. Consequently, the lists of dangerous substances are growing ever longer. Thus the official list of maximum

chemical concentrations in the workplace registered only two substances as carcinogenic at the beginning of the 1970s; today there are
152, and the trend is exponential. To take another example, the
countryside is being developed at a comparable rate. In spite of
economic downturn and drop in population, the proportion of settled
and built-up areas increased in the years up to 1985; for the foreseeable future, no contraction of inhabited areas is to be expected. As a
last example, take the extinction of increasing numbers of plant and
animal species. The annihilation of most species has been concentrated into the period since industrialization began, a minute fraction
of the duration of natural evolutionary processes (Ritter 1987, p.
931).

Destruction and protest are symbolically mediated

This balance-sheet of terror could be arbitrarily extended. Yet the
wealth of scientific observations of the hazards due to science cannot
explain why people protest, either tentatively or (as some believe)
vociferously, against these hazards. The destruction runs its silent
course. Technology does not demonstrate, nor do its repercussions.
Songbirds do not demonstrate (and not only because they have become extinct). Nor have sit-ins or cases of 'Greening to Rule' been
reported of meadows and forests.

The hazards that a chemical or nuclear plant actually and potentially represents cannot be interpreted away (that is, people may die of
leukaemia even if public concern after Chernobyl could be called
hysteria); no amount of technological objectivity concerning the
growing danger entails that the population's disposition to protest
will grow. One might even call this the law of independence of protest
and destruction.

People drive on German motorways at speeds which would lead to
a mass suspension of driving licences in the USA, yet they protest
against the dying forests: it is an appealing thing to do. But the
relation between what one will countenance (to the point of deliberate cultural blindness) on the one hand, where the transportation and
death of a neighbour is concerned, and, on the other, a readiness to
protest over green issues, illustrates at least one thing: the profound
significance of the cultural disposition to perceive, and of cultural
norms. These decide which despoliations are put up with and which
are not, and how acceptance of the unacceptable arises and persists
against a background of unquestioned assumptions.

Why don't we create plastic forests? Neanderthal man is as dead as any dinosaur, and the world does not tear out its hair in lamentation at these losses. Despoiled meadows, fields and lakes – what a market for a forward-looking industrial policy! The replacement of nature: the ideal domestic market, the perfect way out of a sales slump! The Alps are sliding into the valleys? We'll design a new mountain range! The injunction to be masters over the earth thus acquires a connotation of affluent Bible revivalism.

What makes genetics possible, on the one hand – the social-instrumental recombination and reshaping of the various species, which had previously been given to us as 'nature' – is, on the other, rendered mutable by the industrially forced death of 'traditional nature' with all of its faults, which earlier naive and primitive societies had to make do with. Let us create a new, willed, designed, artificial, plastic, social nature, in which cars can drive against flexible treetrunks; and where the genetic enhancement of performance, through the production of children with four arms and hands, is no longer foiled by the 'infantile love of man for his own nature' (Lederberg 1988, pp. 292ff). Certainly resistance will arise here and there, out of narrow-minded property interests and from the romanticization of the past. Similar resistance was put up against the industrial destruction of traditions. That is what the museums are there for. Why don't we build a Noah's Ark museum out of the vanishing traditional nature of past centuries – right next to the museum of regional history and folklore – and then the industrial progress of nature will go full speed ahead.

It is not some mysterious sort of alchemy that gives the protest over certain poisons social resonance. Nor is knowledge of the poison and the danger transmuted into social uproar according to clearly defined rules of political chemistry. Such conceptions are implicit, but highly effective technicist-naturalistic misunderstandings, leading many to think themselves able, merely by pointing out the threat, to trigger off the mechanism of protest. Destruction and protest are isolated from one another by the cultural readiness to accept despoliation. This tolerance of despoliation and hazards, however, wears thin only where people see their way of life jeopardized, in a manner they can both know and interpret, within the horizon of their expectations and valuations. The protest issue must therefore be extricated from technology's self-blockade, and address the injuries to social and cultural conditions and expectations with reference to human experience.[1]

It is only if one inquires into the significance for human coexistence of the presence of poisons and threats that one can comprehend why, for instance, the calming formulae of technology and medical reassurance often have precisely the contrary effect; and why those most gravely affected do not fight most tenaciously against their perceived situation – perhaps so as not also to choke on their anxieties.

To illuminate this central point from another perspective, the dying forests do not contain in themselves the reason for the public attention and concern they receive. Every attempt to deduce social and political protest from an objective, natural-scientific analysis of its urgency (jolting the public, as it were, from its slumbers into an awareness of what is necessary for survival) derives from the prevailing technicist confusion between nature and society. The devastation, i.e. the resulting urgency and distress, ought in many countries to exceed that in Germany. It is more than merely forests that are sick and dying, under the chimneys, next to the refineries and chemical plants, in industrial centres of the Third World. Nor did 'real socialism' exactly distinguish itself by its ecological sensitivity. Many slopes in eastern Europe have already been stripped bare. Nor have our forests been dying only since the uproar (through the mass media) early in the 1980s. For example, the term 'acid rain' was apparently coined in 1872 by the English chemist R. A. Smith. Data concerning the chemical composition of acid rainfall and its devastating effects on vegetation have been systematically collected in Sweden over the past thirty years. There has been one conference after another. It is not only the forests that are dying. Why is there no protest against dying kale, dying flowers, children gasping from smog, mass death on the roads? Why the outbreak of despair over trees in particular? Or, more precisely, over the death of trees *en masse*, in forests, and above all in Germany? Cultural alarm is already finding expression in the pathetic urgency of the formulation 'dying forest', while the euphemism of 'pre-embryonic experiments' is still accepted.

To put it systematically, cultural indignation chooses between matters of the highest 'objective' urgency, and this choice is not guided by the issues themselves, but by cultural symbols and experiences that govern the way people think and act, having their origin in their history and in societal living conditions.[2]

'Who would want to live without the consolation of trees?' The German poet Günter Eich reclaims this horizon of symbolic meaning, which also sustains social resonances and the political protest of the environmental movement. Destruction and protest are symbolically mediated. Symbols that touch a cultural nerve and cause alarm, shattering and making comprehensible the unreality and hyperreality of hazards in everyday life, gain a key significance precisely in the abstractness, imperceptibility and impalpability of the process of devastation kept alive by the advanced industrialism of hazards.

As reality becomes increasingly unreal and inhospitable, one needs to find symbols of the change – pictures in the mass media of dying forests and seals (the latter are also a favourite mascot of tourist resorts and islands), or of the tree withering in our back garden – in order to manufacture culturally the comprehensibility of the incomprehensible. Cultural condensations, magnifying lenses as it were, are required to allow one to extrapolate from the small and everyday to the large, thus bestowing control in everyday life over that which takes its course beyond the horizon of the perceivable and imaginable. In this sense, protest against 'dying forests' is also a rebellion against the cultural threat which still resonates throughout the German-speaking world in the metaphor of the forest, in religion and art, folk songs and *Wanderlust*, from Hieronymus Bosch through the Romantics to the present day. It is a rebellion with democratic undertones, for it reclaims some of what the pan-scientific civilization of hazards has expropriated from human sensibility.

Thus ecological protest is a matter not of natural, but of cultural, fact; a phenomenon of cultural sensibility and of the attentiveness of institutions. Against this background, natural devastation is experienced as cultural and political alarm, and then becomes the subject of social actions and protests.[3] The misapprehension lies in the very topic itself. Nature appears to cry out, and somehow we manage to hear its muted voice.

The death-reflex of normality

Three mutually independent parameters lie between the hazard and the refusal to accept it: the extent of the devastation (how global and systematic it is); public, social knowledge of it (through the various filters of upbringing and the mass media); and the symbolically mediated evaluation of this knowledge in cultural acceptance.

Upon closer scrutiny, information (still very scanty, anecdotal and journalistic) about the behaviour of humans after major accidents and in almost hopelessly hazardous situations – such as life in communities near toxic wastegrounds, or in the vicinity of large-scale hazards – suggests the following hypothesis: as the hazards increase in extent, and the situation is subjectively perceived as hopeless, there is a growing tendency not merely to accept the hazard, but to deny it by every means at one's disposal. One might call this phenomenon, paradoxical only at first glance, the 'death-reflex of normality'. There is a virtually instinctive avoidance, in the face of the greatest possible danger, of living in intolerable contradiction; the shattered constructs of normality are upheld, or even elevated, as if they remained intact.

In their study of the anti-nuclear movement in France, Alain Touraine and his co-authors report that the atomic hazard taken by itself can hardly be made a political issue, by virtue of its very imperceptibility and global character. On the contrary, opposition to it is usually ignited by issues of land ownership and by the collision of a local community with an industrial world controlled from without. The patterns of protest are interchangeable with, say, resistance to a hydroelectric dam (Touraine et al. 1982, pp. 105ff).

In his exposition of major accidents, their outcomes and reactions to and after them, Patrick Lagadec repeatedly comes up against the phenomenon that the people most gravely affected by the hazards are frequently those who seem most determined to repress them. This is also true of the Seveso victims: 'Many did not criticize those who denied the hazards arising from dioxin. On the contrary, they displayed a truly boundless trust in those who played the dangers down, while those who stressed the toxicity of dioxin but knew no appropriate counter-measures met with bitter and occasionally pedantic criticism' (L. Conti, quoted in Lagadec 1987, p. 41).

> The populations's scepticism and indignation were most apparent when the people evacuated from the zone of greatest danger . . . got into their cars on the morning of Sunday 10 October, and returned to Seveso. They broke through the barbed-wire fence and moved back into their homes. For hours the contaminated zone became virtually the theatre of an uncanny spectacle. In the black comedy that ensued, the actors played out their lives in the safe and familiar world of the past, the world prior to the disaster that had left the scene largely intact. The houses and gardens, the grass and the countryside – it all looked so hospitable! People invited each other to dinner or picnics. This 'authentic' spectacle only came to an end when the police and gendarmes were deployed, and when the provincial and regional authorities appeared. (Lagadec 1987, p. 42)

Such clinging to normality becomes understandable if one recognizes that hazard situations do not merely introduce hazards, but turn all living conditions upside down. Aside from, and behind the background of, the unreality of their formulation in natural-scientific-medical terms, hazard situations lead to a maze of extreme social

dependencies, whose unreality is also grounded in the radical contradiction in which they stand to the otherwise prevalent and unquestioned rational-democratic assumptions.

The normal perceptual tools break down, as do all rules of conduct. The extent and universality of the threat constituted by hazards
creates a new kind of primitivism. Hazards bring about cultural
blindness – while the eye still sees, the ear still hears – because our
senses fail us only in respect of chemical and nuclear contamination.
This usurpation of the sense organs always obtains in the scientific
civilization of hazards, but only enters consciousness at the moment
of obvious danger. There is a corresponding authoritarian technocracy of institutional definitions of hazard, at best loosened up or
countered by dissenting voices within the experts' camp, confusion
over precedents and blatant cover-ups, whose only result is to make
anything possible. Thus in culturally purblind everyday life, one no
longer knows which of the (dis-)informing puppet strings, on which
one's life depends, to trust.

Second, hazard forces one to rediscover human beings as natural
entities. Social praxis is predicated upon natural assumptions about
life, whose social organization one becomes conscious of only when
those assumptions are endangered. To reduce it to a truism: there can
be no social praxis without breathing. To the best of my knowledge,
breathing is yet to be discovered as a social activity by sociologists
and social theory. For human nature, as adopted into the civilization–
nature context, is structured by its compulsory integration: its
decisions are closed, bypassing everything the social sciences ever had
to say about norms, consensus, decision, power – or, rather, turning
them into their converse. Breathing integrates a human being into the
natural despoliation that he perfects through industrial means. Social
praxis is undercut, via social organization and injury to its basis in
human nature, through an archaic form of dependence which knows
modernity only as the negation of itself. It is precisely the highly
developed industrial systems which render impossible any escape
from the collective self-jeopardization of the natural basis, in human
nature, of social action. One might at best install air filters in homes
and legally proscribe the breathing of unfiltered air from outdoors; or
elegantly circumvent the clumsy solution of using gas masks by means
of artificial nose and throat implants – as a technological waste-
disposal gadget for the 'air consumer', his consumption astonishingly
enhanced and his original body no longer usable. If (for whatever
reasons) this should seem too expensive, there remains only the social
construction of non-toxicity. It does not, admittedly, inhibit the

effect, but only its designation. We swallow it, therefore it isn't toxic – this is the last way out in view of the presumed dead end of universalized contamination.

Of course, it is possible to react 'sensibly' to civilization's fated dependence on nature, i.e. to doubt the fact of poisoning, to counter the effects of data by means of contradictory data. That might be a momentary consolation, but it is no help against poisoning.

The desire or need to breathe, eat, drink, etc., which humans share with animals and plants, is becoming a gate that can no longer be barred against hazards; only symbolic barriers, namely maximum pollution levels, can be erected against them. It may be possible to select one's food, and thus privately attempt to elude the creeping universalization of hazards; one could also try to pay in hard cash for unpolluted air. Thus private efforts to compensate for hazards would bring about new forms of social inequality, measured in terms of filters, opportunities for foreign holidays and, above all, health and nutritional standards. These could be interpreted as a kind of expensive, private counterbalance, available to the individualistic educated and propertied classes, to the universal hazard that also threatens them. Yet these bolt-holes become less and less available to fewer and fewer people, while the industrially perfected normalization and universalization of harmful and toxic substances continues. Moreover, we are always dependent on knowledge; we must know what to avoid. Yet this knowledge is forever running away from itself. In the borderline case, the meaning of this for everybody is that the very thing which integrates them into the now socialized, natural context of life delivers them helpless to the industrially manufactured and normalized dangers.

With that, however, symbolic detoxifications gain significance as a last illusory salvation – as straws to clutch at, as it were, though far out of reach. Maximum pollution levels can be dealt with politically, and they systematically launder poisons. For that very reason, these poisons can be disseminated as they are, without let or hindrance. Involuntarily, though, in the face of the rising levels of pollution and contamination, the very stringency of maximum pollution levels brings out the fact that it is supposed to expunge from consciousness: namely, that in the nature–danger context that advanced industrialism has become, social action in its residual sense of escape, exclusion, rejection, is abolished. Humans are meanwhile integrated into its contrary, i.e. a collectively threatened existence mediated by the nature–society context, that modernity had promised under the aegis of autonomy, choice, individualism.

It is precisely the knowledge of the hazards, hailed by many as an escape route, that renders everyday life so hopelessly dependent, and puts it in the most extreme antithesis to the democracy that otherwise obtains and is proclaimed. We have here a kind of negative socialization predicated essentially on dependence, the dependence of a blinded everyday culture on the hazards that elude its judgement, its decisions and frequently also the chance of private escape. The situation is extreme in several respects. It is impossible to consent to this state of affairs. Everything we had known about the integration of standards is thereby rendered obsolete. This leads, on the one hand, to rebelliousness, where one externalizes the real madness to which one sees oneself exposed: thus in estates near toxic waste sites, well-behaved citizens, normally loath to make a fuss of any kind, are transformed. Unrecognizable even to themselves, they take up arms against a 'dictatorship of actuality' they cannot break. At the same time it produces a sense of hopelessness that comes with a dawning consciousness of hazards and forms the basis of a forced pledge of faith in the security of the old order. People believe with all the depth of their being in the need to survive, and they believe those who permit them to restore and secure normality in the act of putting their faith in it. Faith, if need be, is a sword to be wielded as powerfully as possible against all those who question it – because only thus can the chaos that threatens outside one's front door be excluded. In other words, staring into the abyss of dangers becomes integrated into normality.

This option is always available and obvious, as the consciousness of hazards is predicated on a kind of self-expropriation of the senses. It is precisely the imperceptibility of the dangers that appears to turn the re-establishment of normality into an individual choice. One need only trust one's own senses, throw aside all extraneous data, and the spectre appears to vanish into the unreality whence it came. Paradoxically, it is faith in science – which in Germany is frequently taken to the point of superstition – which prevents the re-establishment of this basic right to believe what one perceives in a kind of private *coup*, through re-empowerment of the senses. This back-to-normality movement has every expectation of winning followers from the free-fall into hopelessness. It is now certainly hindered by what used to ensure normality: lack of faith in one's own perceptions in the civilized world, and dependence on science, which has long since cut individual judgement loose from its anchoring in individual experience.

Dangers, even where they remain impossible to calculate, involve greatly increased labour for housewives and mothers. In order to fulfil their own expectations, they must now privately clear away the socially created hazards that break in on all sides. Furthermore, the failure of the institutions that permit and normalize the poisons threatens to turn into the personal guilt of the mother. Putting her faith in some data or other, she has ultimately put her loved ones in the very danger she had intended to prevent. No wonder women are preoccupied with poisons at night, despite the suspicions of jealous husbands (Kerner 1987).

The German 'anxiety' miracle

If destruction itself does not spark off protest, if hazard situations resist perception, then is it not actually a miracle that in Germany, rich and well-fed by world standards, the overwhelming majority of the population should have been alarmed by environmental issues, irrespective of the parties they support?[4] Why is there ecological protest in Germany? How can we explain this 'miracle of anxiety'? Does protest against the rape of the environment also express cultural insecurities and injuries? One can see three historical phases mirrored in the 'nature politics' of the ecological protest movement.

First, the rat-race of advanced industrial society tends to encourage people in their search and yearning for its antithesis, yet the latter is worn down systematically in the maelstrom of modernization. The cultural model of nature which both emerges under the living conditions and highly developed industrialism of Germany, and is threatened by them, is one of calm. Internal and external experiences thus harmonize in the act of protest against natural despoliation, and condense into an awareness of being endangered, where the ghostly, barren mountain slopes and the exploding nuclear and chemical plants only give external confirmation of the inward experience.

Second, this inward experience is that of an inflationary devaluation of the guiding assumptions and utopias of class, family, wife, husband, marriage, parenthood, occupation, etc. (Beck 1992, part II). In the wake of this radicalized problematization of all principles for conducting one's life, nature appears as a passageway to 'consecrated' self-evident truths; as an endangered store of unbreakable rules to be discovered, guarded and cultivated.

Third, the awakening protest expresses the shattering of constructs that have hitherto made environmental hazards what they are, i.e. hazards whose effects are not felt in the social institutions themselves, but which can at once be tamed and compelled into the semi-irrelevance of the environment. In fact, 'natural despoliations' are all

about preventing domestic social unrest over prices, sales opportunities, property rights, failures of policy, votes, legal culpability and
damages, power displacements. This (and herein lies the appeasement
in the consensual formulation of an 'ecological crisis') remains concealed and distorted in the struggle over air pollution, ground water
pollution, protection of the species. Talk of 'environmental problems'
is thus an unstable compromise between the veiling and unveiling of
social interests and responsibilities: it is a kind of 'pre-emptive strike'
by a profoundly insecure industrial society whose interests and power
have been challenged – a pre-emptive strike in the no man's land of a
nature that no longer exists.

Ecological protest does not ignite amid the social milieux in greatest danger, where poverty, filth, noise and risks have formed an
unbreakable alliance of threats; it is ignited among the middle-income
majority, whose standards of safety and of health were cultivated and
developed during the golden years of the 1960s, on the basis of a
(modest) share of affluence and property, knowledge and education.
This majority now sees itself robbed by ecological despoliation of the
fruits of its labours – leisure, house and garden. It is not the despoliation of nature, but the jeopardization of a specific cultural model of
nature (whose contents and endangerment indicate the level of development of wealth production in Germany) that provides the sound-
ing-board for the ecological alarm of an entire society. The image of
nature destroyed, and whose destruction is experienced, is the counter-image of the hectically mobile, meritocratic, affluent society; and
the latter jeopardizes the enjoyment of what has been achieved by all
the tools of progress: cars, roads, consumption, mobility.

Stress increases with gross domestic product, and with it the urge
to drop out of the rat-race – in the evening, at the weekend, on
holidays or in middle age. Wherever one attempts to flee from stress,
one does so through its own tools: the leisure and tourist industries,
motorways, ski-lifts, etc. These assimilate stress itself into the last
'nature' (for which read 'stress-free') reserves. Thus, in the process of
fulfilling this wish and of providing the material conditions for its
fulfilment, increasing numbers of people are seeing it undermined and
buried.

Overbred industrialism develops a dialectic that ultimately poisons
the pleasure even of its beneficiaries. This is, so to speak, the external
correlate of the inner experience, which is becoming widespread in
the affluent society of developed capitalism: life becomes more hectic,
insecure, self-destructive – in life-companionships, careers, the search
for peace and happiness. The lake one was about to leap into is

revealed as a sewer, the superb, crispy lettuce in one's mouth turns out to be contaminated and foul. The dissolution of inner worlds – the painful shattering in everyday life of previously unspoken assumptions between husband and wife, the dissolution of corporate class cultures, the spiralling mobility of neighbours and acquaintances, the unpredictability of political life, the fading of utopias, and the perpetual uncertainty and explosiveness of even the most harmless questions of detail – this loss of inner stability in the conduct of one's life finds its correlative reinforcement in the external hazards of nature and industrialism (cf. Lau 1985). The fatalism of a self-destructive dynamic of progress corresponds to the fatalism with which the disintegration of 'havens of privacy' actively runs its course and is passively tolerated. The inaccessibility to experience of environmental destruction finds its imagery in inner endangerment, and both condense into an experience of a catastrophic age that veers contradictorily between fatalism and protest.

The ecological movement is not an environmental movement but a social, inward movement which utilizes 'nature' as a parameter for certain questions. These owe the attention accorded them to everyone's growing horror at the fact that what it had been supposed we could foresee has taken root in a foreseeably unforeseeable jeopardization, not only of nature, but also of leisure, recreation, health, life, capital, property, performance.

Talk of 'natural despoliation' also relieves one of the burden of moral justification. Scientific eco-networks achieve more than the rag-bag of unsupported moralizations ever managed: the impalpable appears to become palpable, the ambivalent unequivocal, the groundless susceptible of proof. Chemistry ('maximum safe levels') prevails – perhaps with a drop of alchemy – where moral philosophy formerly spread its uncertainties. Now one can argue about SO_2 contents and the link between smog and death rates, instead of the interpretation of the categorical imperative.

Natural despoliations are the morality beyond morality. They permit a critique of society beyond social critique – with all of its double Dutch and murky Marxism. The ecological critique is the most powerful brake that can be applied to industrial momentum. Developed technocracy barely gives those it affects room to breathe or to argue, and in the (last?) twitch of protest people put together the following message: whatever is destructive must be switched off, put into reverse. 'Natural despoliation' also links the general to the particular, in that it permits the foundation of an 'alternative lifestyle' unsupported by any foundations. 'Nature' is the testable form of a concrete utopia, beyond the province of utopias riddled with doubt and turned into their contrary. It is a way of normalizing life while pretending to live in freedom from norms – a way, that is, not of posing but of answering the question of how we intend to live.

This makes it understandable why the environmental movement (leaving aside for the moment the farmers directly affected) is recruited especially from milieux and groups where disenchantment and the loss of unquestioned assumptions have reached an advanced stage.[5]

There is no end yet in sight to these destructive changes. In the confused diversity that emerges and shifts here, there are also signs that the prevalent conception of man is itself correspondingly damaged. The deluge of violence, destruction and disasters satisfies a (no longer merely latent) sadism of the epoch. Is the 'good person' now only a false label, under which one is conventionally enjoined to pass off goods that are already beginning to breed the contrary?

A couple of snapshots: executions take place daily on the television screens of our living rooms, to the accompaniment of delicious horror and fingers rooting in crisp packets. Disaster tourism: granny holding an infant by the hand, in front of destroyed buildings and burned-out cars. Let us take another little detour to the next saloon bar of 'silent horror'. The tourist guide to disasters with one-, two-, or three-star ratings, tried and tested by our creepshow specialist.

Industrial fatalism: the internal jeopardization of one's social habitat prepares one to accept the devastation outside. The wild affray of cars in the traffic jungle prepares one for major accidents, which one then 'sits out' just like the motorway tailbacks at the start of the holidays. The new dogmatism: strained 'destruction hypotheses' borrowed from other fields of knowledge are divested of their uncertainty, and superelevated into hard facts which 'prove' the madness of the age. The doom-laden atmosphere and ecological protest in affluent Germany are, in my opinion, to be explained by its very wealth. Industrialization means wealth, but also proximity to industrial hazards: air pollution, the break-up of communities, the spread of concrete jungles, perpetual mobility, thereby endangering 'nature' (in the sense of peace, recovery) from within and without. The need for nature and its destruction are mutually reinforcing, and are not suppressed by other worries (war, famine). High levels of safety and safety consciousness, the density of bureaucratic regulations and raised standards of health and hygiene combine with mass-media dissemination of the growing destruction. The rise of the scientific attitude has transformed our conception of risks and their consequences. All these things are indicators of wealth, permitting the emergence of sensitivity to and consciousness of survival issues in the high-density industrial oasis of Germany, with its high living standards and fanatical regard for cleanliness and order.[6]

Many parts of the world, especially in the West, are still spared the devastation, damage, poverty, acute distress, hopelessness, misery, poverty of expression and not least the wars, which are more or less a part of everyday life nearly everywhere else in the world. Against this background, consciousness of hazards and the declining acceptance of ecological devastation are the reflex of a privileged existence in a niche protected at great expense (Elmar Koenen, personal communication). Knowledge of the disastrous world situation, which cannot be entirely held down even here, throws a shadow of dread and hope on those metropolitan islands where a (privileged) consciousness of threats from the society of global risk first emerges.

3

Industrial Fatalism: *Organized Irresponsibility*

If there is anything that produces unity across the entrenched political divides, then it is the conviction that we are imprisoned by our dependence on rationality; that productive forces give rise to the liberation that enslaves us; that we are captives of a reason that threatens to turn into its contrary; or that we differentiate by function, to the point where everything always becomes more functional and more differentiated. These are all theoretical variations on the basic experience of the age, that of industrial fatalism. Variations in a major or minor key are scored for versions set in heaven or hell, for flute or drums, from Comte to Adorno, Marx to Luhmann, to name but a few.

Walter Benjamin gave an impressive literary testimony to the fatalism of progress.

A Klee painting named *Angelus Novus* shows an angel looking as though he is about to move away from something he is fixedly contemplating. His eyes are staring, his mouth is open, his wings are spread. This is how one pictures the angel of history. His face is turned towards the past. Where we perceive a chain of events, he sees one single catastrophe which keeps piling wreckage upon wreckage and hurls it in front of his feet. The angel would like to stay, awaken the dead, and make whole what has been smashed. But a storm is blowing from paradise; it has got caught in his wings with such violence that the angel can no longer close them. This storm irresistibly propels him into the future to which his back is turned, while the pile of debris before him grows skyward. This storm is what we call progress. (Benjamin 1978, pp. 257–8)

The politics of aporia

The questions and criticisms have been backed up by experience; they are becoming more vocal, growing beyond a mere readiness to sign appeals. The word 'no' is becoming a popular word with the Catholic Church and the feminist movement among others. Switzerland and the trade unions are giving it consideration. Discontent is growing across party lines, and even within some government ministries. Rainbow coalitions are coming into sight, ranging from angry farmers to medical critics, geneticists and lawyers. But the process at issue is taking its inexorable course, throughout the world. It is precisely the activists, those ready to protest, who see themselves condemned to a politics of aporia. When every good argument, every new approach, provokes the mouthing of ten 'hard' facts implying the contrary, political commitment almost inevitably turns into fatalism.

Suppose the impossible were to come true. Are there not another couple of hundred nuclear power stations 'outside our front door', in Europe, in the world? Granted, the bio-industrial investments have been activated. But would not putting the brakes on 'commercial eugenics' (Rifkin 1985), the Second Creation of genetic engineering, rob our agriculture of its international competitiveness anyway? No sooner has a vanguard of determined protesters united against human-embryo experiments than the cry goes up, 'You are preventing life-saving Aids research!' Freedom of research, the unforeseeability of the consequences of scientific work, is cited as an internal law against the imposition of controls. Besides, does not sympathy for the millions of unemployed compel one to the practised gesture of looking away? Laws? Aren't there more than enough? Must the rate of convictions, already notoriously low, drop even further?

Futile wrestling bouts are staged on European political platforms about such issues as the crucial importance for our survival of reducing maximum pollution levels, or of taking measures to halt the disappearance of species from the North and Baltic Seas. Time and again these merely demonstrate how the free passage of poisons under the prevalent system of norms is further ensured by the international self-blockades of a preventive industrial policy. Thus the government successfully transforms its near-inaction on the international stage into activity, basking in the glory of appearing to crack the whip.

The controversies, by contrast, take place in arenas far from the centre of power; and, if only for this reason, the uproar (which is fitful enough) remains nearly without consequences.

Secondary consequences: these do not merely presuppose constant causes and perpetrators, but are also a stab in the dark of the future, which is anyway uncertain and unknowable.

Ethics: idle oaths sworn after the deed, the more profitably because they are genuine only in so far as they cannot be guaranteed; and because the dramaturgy of ethics implies the dramaturgy of its antitheses, in which it goes under.

Laws: these constitute an assent, incomprehensibly condensed into paragraphs and authorities, to everything new in science, economics and technology. The pruning of details legitimates the main trend. There is no better way of symbolically detoxifying the reality of the danger, which for precisely that reason is rendered irreversible.

Accidents: the judgements passed follow rituals that permit catastrophes to sink to the level of normality, by concentrating on extreme cases.

Public opinion: not everything stirred up here runs its course in the circumscribed semi-irrelevance of the 'media circus'. It is also subject to its own laws: as statistics show, the 'half-life' of public alarm fluctuates from a couple of weeks to several months, depending on the type of disaster.

Social movements: taken literally for once, these mean coming and going. Above all, going. Self-dissolution leads the way.

The Kafkaesque experience of protest

To argue against science over the matter of hazards, one must deploy science's own tools. Like arm-wrestling against oneself, one has to pull oneself over the table, before one can even enter the arena. That is, protest must accept the basic assumptions it intends to contest before it can even utter a word.

The observable consequence is that critics frequently argue more scientifically than the natural scientists they dispute against. The struggle for efficacy compels them to do so. One attempts to play off results, methods and authorities against each other, and so falls prey to a naive realism about definitions of the dangers one consumes. On the one hand, this naive realism of hazards is (apparently) necessary as an expression of outrage and a motor of protest; on the other, it is its Achilles' heel. It is a variant of the 'naturalistic misunderstanding' (chapter 2), mislaying opportunities for action, exposing protest to self-refutation from inside and outside as it develops its technicist premises, and thus delivering it up to the experience of industrial fatalism.

This self-imposed handicap of naive realism about hazards extends further than their relationship with technology and science. The same can be shown to hold true of one's identification of the enemy. Anyone who believes that the 'causes' to be brought to book are situated where the filth pours out of the chimneys has not understood the protective barriers of definition that companies, with the support of the law and of science, have erected around themselves. The deed and proof of it are worlds apart – worlds which, under the prevailing relations of definition, foist the burden of proof on people with less chance of success than they have of filling in all the draws on their pools coupons (see also chapter 6).

If, however, contrary to expectations, some minor, immanent contradictions can be blown up to the point where experts go on the defensive, it turns out that one has not been fighting against the responsible or guilty people. What say do scientists have in industrial and government hierarchies, and for what are they individually responsible? Certainly, they are the first people one has to deal with in court; judges are required by law to judge according to the prevailing orthodoxy of the specialists (cf. p. 133). Yet, if the protest continues, there remains the wide gap between the undertaking and its fulfilment; co-operation between the controllers and the controlled; the impenetrable grey area between tolerance and negotiation. Even if by a miracle something can be proved here, protest finally loses its way in the maze of information carefully erected by the companies, from which hardly anyone emerges who entered actively protesting.

Thus what is at issue is an elaborate labyrinth designed according to principles, not of non-liability of irresponsibility, but of simultaneous liability and unaccountability: more precisely, liability as unaccountability, or organized irresponsibility.

People are still wondering what happened to the horror, the shock of Chernobyl. Perhaps the answer is easy to find. Perhaps we western Europeans have caught up with an experience that has long been a part of everyday life in eastern Europe: the reality of Kafka's *The Trial.*

Life and praxis in the risk society have become Kafkaesque in the strict sense of the word – if this concept designates the absurd situations available in real life to the individual in a totalitarian, labyrinthine world that is opaque to himself; situations not be characterized by any other word, to which neither political science nor sociology nor psychology can provide a key. What Max Weber develops from the point of view of administration as 'bureaucratic rule', and what Hannah Arendt interprets, in the extreme case of the bureaucratically planned and executed annihilation of the Jews, as the 'banality of evil', Franz Kafka exposes from the perspective of an individual victim who, as Milan Kundera puts it, 'is a child gone astray in the forests of symbols' (1988, p. 63).

It begins, as in Joseph K.'s case, with an invasion of privacy – two gentlemen arrest him while he is in bed. The hazards slip past every boundary, past all the constructs of a 'life of one's own' that we hold so dear; they are simply there, impervious to any decision one might make, they penetrate the most intimate areas of our lives. The whole canon of everyday knowledge breaks down. The instruments of defence – our senses, our faculty of judgement – have been expropriated overnight. But whoever mistrusts his own senses is 'arrested', hangs from the puppet strings of centralized data and labyrinthine data administration systems (the mass media, ministries, experts and so forth), of which he can only say for certain that they are contradictory, and therefore deceive him.

The further one goes down the branching corridors, rooms and subterranean tunnels of hazard manufacture and administration, i.e. the lawcourt, the more apparent it becomes that here is a great bureaucracy, extending into and out of the factories themselves; it consists of general recklessness and uncertainty, growing ignorance or half-baked knowledge, falsification, cover-ups, etc., under the truly inspired organizational principle of unaccountable non-liability.[1]

Yet this whole comedy of the absurd (Kafka himself and his friends could not stop laughing during the readings in their little circle) breaks out only when there is occasion for protest. Without expert judgement the critique is hollow, and with expert judgement it is – refuted. Protest must speak the language of a science that serves as much to bring about the hazards protested against as it serves the cause of protest itself, and of those who counteract it.

As it passes from one court of appeal to the next, protest accordingly meets the scientific reception parties whose defensive welcome is as deft as it is effective: they greet protest with the sweetly duplicitous question of what precisely is the cause for concern (though they understand it only too well and, indeed, essentially share it) and then lead it by the burden of proof to which it has shackled itself upon entering, by the ring through its own naive nose, into the labyrinths of provable unprovabilities. Here, during the stage when they administer the risks of their own creation, the sciences are in their true element. They lead protest – not astray, God forbid – but into the mysteries of rigorous science; which, if protest is not to convict itself of irrationality pure and simple, must also be its own most deeply characteristic, innermost disposition. Whoever takes this advice and really thinks things over, can ultimately only beg to be forgiven for entering so brashly – and, moreover, seek in himself the causes of the tormenting consciousness of hazards that has goaded him on: exaggerated fears, subjective failure, masked insecurities projected onto the environment as a result of joblessness, marriage, family, etc.

The way in which protest chokes on itself resembles the spiritual journey of Joseph K. to the last detail. The absurdity of the punishment, i.e. the threat from hazards, is so impalpable, so unbearable that in order to find peace the accused seeks in himself a justification for the punishment. The trial begins with submission as a principle of resistance, and ends with the accused accusing himself on his accusers' behalf – and then delivering a verdict of 'guilty'. It will surely not be long before the slight increase in miscarriages in Germany after Chernobyl, not statistically provable anyway, is traced back in expensive large-scale researches to the whipped-up anxieties of mothers, and nothing else. Whence it follows that mothers who take this to heart will have nothing to fear in future but fear itself.

Organized irresponsibility

Extricated from the literary analogy, that means consent or revolt: the threat that straddles the categories of the institutions that administer it, does not exist because it is administered, and grows because it does not exist.

Thus Chernobyl has taught us at least three lessons. First, that the worst-case scenario is possible and real, and that probabilistic safety is deceptive; second, that the abolition of nuclear power has become an accepted political possibility; third, that the amateurish mélange of state and technological authority has given way to a near-perfect procedure. A flexible implementation of maximum pollution levels; compensation; a technically oriented legal procedure which downplays the hazards; the administrative blockades of overzealous abolitionist policies; systematically celebrated symbolic detoxification in the mass media; data centralization – all these mean that the much-lamented 'worst database scenario' has ultimately proved to be a highly successful policy of absorption, from which lessons have surely been learned for future cases.

Yet no matter how perfect the normalization policy, the institutions have evidently been shaken: the hasty repudiation and assignment of guilt; the disintegration of expert certainties; the horror at
our collective ignorance in the face of a hazard as invisible as it is
universal; the collapse of entire industries, market realignments;
snowballing compensation payments and so forth. The crumbling-
away of the façades of normality under the harsh spotlights of the
mass media, on the one hand, and the speed with which these
breaches in the overall structure were mended, spatchcocked together
under the gravitational force of normality, on the other; taken
together, these clearly indicate the central contradiction to be illuminated in what follows: namely, the contradiction between hazards
produced by and within the system, and those unattributable to the
system, which is neither responsible for them nor capable of dealing
with them.

In everyday life, as in politics, the economy and even the sciences,
it is naively assumed that hazards originating in industrial-technological, scientific-economic development (and for which no god, devil or
demon can be blamed) can also be exposed, tracked down and dealt
with – and, with sufficient resolve, avoided – by the customary
criteria of causality and guilt. Yet this view is one of the naive
attitudes which cover up the system of organized irresponsibility. For
it is precisely the other way around: it is the application of prevalent
norms that guarantees the non-attributability of systemic hazards:
hazards are writ small as risks, compared away and legally and
scientifically normalized into improbable 'residual risks', making
possible the stigmatization of protest as outbreaks of 'irrationality'.
Those who uphold maximum pollution levels turn white into black,
danger into normality, by act of government. Whoever waves the
banners of rigorous causal proof while demanding that the injured
parties do the same, not only demands the unachievable, as science
has meanwhile adequately confirmed, but thereby also holds aloft a
shining shield to keep rising, collectively conditioned hazards out of
the reach of politics or attribution to individuals.

In other words, the institutions involved and affected do not
merely dispose of highly effective instruments and strategies for 'normalizing' industrial self-jeopardization. This normalization has been
achieved precisely by demanding and doing the same as has always
been demanded and been done: inquiring into 'causes' and prosecuting 'guilty parties' in accordance with hitherto valid conceptions. The
greater the shock with which public consciousness apprehends industrial self-jeopardization, the more important it becomes not to allow

what is produced under systemic constraints to appear as such – by using every means available to whip up a froth of demonstrative action.

The superficial political solution to this central contradiction of a self-endangering society is to propagate and safeguard an industrial fatalism in which the products of the system are not (attributably) due to the system; and in which the culpability, which can no longer be palmed off on the external world, is supposed to derive from an internally created, ineluctable natural destiny of civilization. The weather (smog) is no less a component of this than the application of the otiose rules of attribution (causality, guilt) that prevail, as well as the preservation of prevalent, unequal relations of definition (burdens of proof).

As for experiences, orientations and ideologies, three types of consent to the industrial dynamic are discernible – one positive, one negative and one cynical.

Positive, negative and cynical fatalism

Faith in progress[2] is certainly still the dominant (I was about to say 'healthiest') attitude, because it affirms what happens anyway. For this reason it does not exactly exude the perfume of freedom. By its consent, positive fatalism covers up the fact that industrialism has put a new coercive system in the place of the old one. Faith in progress is genuflection, as it were, before a throne that can no longer be abolished, whose power cannot be diminished or made worthy of consent through an exchange or separation of powers; whose power is not even attributable to one person or authority. This self-created anarchy (*Niemandsherrschaft*) of progress, as Hannah Arendt describes it, is so impenetrable that the only thing left is to assent. One must love a tyrant one cannot overthrow or vote out.

Faith in progress maintains this wisdom of welcoming the inevitable, for its very inevitability. In view of the superior force of circumstance, it is unable to act, and yet in the age of democracy and enlightenment it is actually obliged to act. It resolves this contradiction by accepting it in the light of higher values – namely, those of the progressiveness of what happens anyway. If one toes the line now, one does so in the name of the illustrious project of perfecting mankind, and not from the bare want of alternatives. The affirmation finds the strength here and there to trim the risks, to avoid the very worst of possible cases – or at least to hope to.

The pessimists, the clairvoyants of doom, the negative fatalists, have certainly won followers. Perhaps there has even been desertion behind the scenes from the ranks of naive industrial clan consciousness. If recent investigations into youth attitudes can be taken to confirm this, then the fraction of stalwart fatalists of progress has suffered a severe decline in numbers.[3]

It is also interesting that many desert directly from optimism to cynicism, skirting pessimism. There are reasons for that. Pessimism is extremely uncomfortable, in contradistinction to cynicism, which rejects every evaluation including the negative one, partly from insight into the latter's fragility and distortedness, partly from pragmatism.

Cynicism about progress allows one to live comfortably once again. It lays down the burden of defending a now unstoppable naive industrialism, or of taking up arms against it. One can recline at one's ease, or dance on the rim of the volcano: cynicism lends the post-consumerist consumer rush that touch of absurdity, the frisson of panic. Refusal, which is consummated in the refusal to refuse, constitutes a near-impenetrable padding behind which one can combine enjoyment of the comforts of consent with the advantages of refusal – in the knowledge of always being at the forefront of the age.

On the other side there are those who see enlightenment converted into its opposite, as a train heading in the wrong direction, driverless; they can neither get off, nor pull the emergency brake, nor make the train run backwards. They run to the rear of the train and press their noses against the rear window, but they're moving. They hammer against the windows with their fists, but they're on their way. And they themselves know that they're on their way, and that none of their gestures – no matter how thoroughgoing, perspicuous, artful or negative the critique – can ever stop the train or reverse its direction. Not because this possibility is excluded in principle, but because their radical critique feeds on this inexorability.

The other travellers owe a great deal to this little group of critical sceptics and despairing critics, who do not take their ease on the cushions of cynicism. For at least they feel the lack of what hardly anyone else will admit to missing: braking devices, steering systems or the like, a promise redeemed – democracy, self-determination, enlightenment. They are the ones whose sad, furious, ludicrous gestures at least still remind the travellers of the absence of these elementary requirements.

Over and above all the contradictions, it is nonetheless true that optimism, pessimism and cynicism completely agree in their diagnoses of the industrial dynamic. This is clearly true of the optimists of progress. But even the pessimists ground their pessimism precisely in the inexorability and uncontrollability of industrialism. In other words, there is an inflationary critique of faith in progress, but none of industrial fatalism. Enlightenment seems to end up as the fatalism of developed industrial society, making everything achievable with one hand, while the other hand consecrates its powerlessness in the baptismal waters of progress. Only when protest takes direct action does it often have enough thrust initially to elude this background consensus between the antagonists – the insight into the industrial dynamic which paralyses action and relieves one of the duty to act.

How industrialism annuls its own premises

I use the term 'risk society' for those societies that are confronted by the challenges of the self-created possibility, hidden at first, then increasingly apparent, of the self-destruction of all life on this earth. At this point I wish to distinguish between two phases: a first phase, during which we live under the threat of self-destruction, but think and act in the categories of industrial society – scientifically, legally, economically, politically. At first this is surely unintentional, enabling previously unquestioned assumptions to persist; but it becomes increasingly strategic, because the confusion of centuries appears to lend stability to the old order against the onslaught of the new.

In the process, a technocratic variant of the risk society appears, wherein the systemically conditioned insecurities are perfected into a technocratic road to power (see chapter 7). The risks and disasters are preliminary exercises towards a culture of wardship. The plausibility of experts is only superficially undermined by accidents, and essentially serves to expand the power of industrial-technocratic elites. In this developmental variant, the backlash of public opinion after Chernobyl was, as it were, the last twitch of democracy after its demise.

The second phase develops as the risk society mechanism, which had been repressed and realized in its suppression, comes into play and is acknowledged; and also as the manufacture of attributability and liability, under the system of non-guidance and non-liability which emerged with industrialism, becomes the central problem of political development. This transition must essentially be enacted in the domain of knowledge, in reclaiming the ability to learn; thus it is by no means a deterministic process, and occurs only once ecological hazards are recognized as forms of social self-jeopardization, and are cut back and contained by institutional innovations. Hazards to plants and animals, to water and human health, become hazards to property and economic performance, i.e. immanent hazards. The end of ecology is the beginning of a political ecology which comprehends the transformation of universalized hazards to life in terms of the systems of economy, science, law, etc. (see part II).

Everything has been conjured up to avert hazards: reason, commissions of ethical inquiry, parliaments, charismatically gifted leaders, spiralling costs and the collapse of nature. Yet these surely form the bass counterpoint to the onward march of industrial development, to whose doom-laden fanfares it has won all its victories until now. Industrialism, having arisen out of good intentions, is armourplated against their bare hands, no matter how frequently or loudly they are invoked. Yet no thought was given to the possibility that progress could be slowed down, interrupted or at least called into question by the victory of progress itself. Realization as annulment: the real challenge and refutation to industrial fatalism comes from industrial fatalism itself (see chapter 4). Against its own will, its planned actions, the specialist treadmill, etc., modernism itself undermines modernization. This, however, leads to social restratifications, power shifts, new lines of social conflict and possible coalitions; ethics, public life, the mass media, new ways of thinking, the actions of individuals and of social movements, win their historic opportunity.

More concretely: the (positivistic) scientific ideal is refuted, not by critical theory, but by the development of an objectified science. Striding forth along the path of specialization, it does not, cannot, stop at the deconsecration of its own foundations and products. Anyone who lets his doubts gnaw at his own fundamental principles can, if he does the job well, realize that he has none. Thus experimental safeguards, which form the very core of natural science, have in the twentieth century quite simply been thought away, doubted out of existence, in long, thorough debates intended to decipher rationality. All this has been successfully accomplished by scientific claims. Ultimately, doubt is victorious over the understanding that owes it everything, and thereby creates a different situation, as yet uncomprehended, beyond 'positivism' and the 'dialectic of the Enlightenment' (cf. chapter 5).

The law suppresses the justice it was supposed to establish. Through a series of legal and procedural technicalities, the lawyer becomes technology's legal adviser. The greater the hazard to basic individual rights, the less the legal protection (cf. chapter 6).

The economy loftily blames the hazards of its own manufacture on the environment, where nothing is attributable. In so far as it succeeds in doing so, it creates and foments contradictions in its own camp. Perhaps it even skates off into a new social structure, whose consequences and conflicts remain unforeseeable. That which is contaminated as environment by, say, the chemical industry, is supplied to the marketplace by another sector: tourism, say, or agriculture, or the fishing industry (see chapter 6).

If this premise proves at all tenable, then the repercussions on lines of conflict in labour disputes will be profound (see chapter 6). If the logic of risk definitions take centre stage, these lines of conflict will no longer be bound by the criterion of non-ownership of the means of production, but by membership of those economic sectors that profit and lose respectively by risk.

If industrial fatalism is true, then it is outdated. We do not intend here to hit it with the truncheon of moral criticism, but will try to lead it, by its own truth, off the beaten track of history: the side-effects of side-effects add up to the picture of a different modernity. The overwhelming feature of the age is not physical – the threat of annihilation – but social: the fundamental and scandalous way in which the institutions, almost without exception, fail it.

Gradually, accident by accident, the logic of institutional negligence turns into its opposite: what do probabilistic safeguards – and thus the entire natural-scientific analysis – have to contribute to the assessment of a worst-case scenario whose occurrence leaves the theories intact, but annihilates life?

At some point the question presents itself: what good is a legal system that prosecutes technically manageable small risks, but legalizes large-scale hazards on the strength of its authority, foisting them on everyone, including even those multitudes who resist them?

How will faith in 'economic growth' and 'affluence' survive, if property and the economic basis of entire industries are systematically run down and annihilated behind the ever-thinning walls of 'environmental destruction'?

How is one to maintain a political authority which has to confront the swelling consciousness of hazards with loud safety assertions, but for that very reason slips into a barrage of perpetual accusations, and gambles the whole of its plausibility on every accident or hint of one?

PART II
Antidotes

4

The Self-refutation of Bureaucracy: *The Victory of Industrialism over Itself*

This book disputes the idea held both in the twentieth century, with its consciousness of itself as the era of disasters, and in the eighteenth and nineteenth centuries, with their faith in progress. Left, right or green, everyone entirely agrees in diagnosing the causal autonomy of industrial capitalism, which according to the traditional schemata augurs either progress or class struggle, but now appears to have turned into the threat to natural species, the ecological self-jeopardization of all life on this earth. This project, not to be taken at all seriously by a sober, level-headed scientist, is bound to be asked how it intends to avoid failing like the sequence of earlier attempts.

Industrialism's causal autonomy may be a hopeless topic or an exhausted one, the theme of the century or no theme at all. If we cower before it, continuing to enforce and impose a taboo on this living mechanism, a product of hands, brains, laws, power, the economy, science, habit, indifference, then the words with which we all flatter ourselves – words like democracy, responsibility, rationality, freedom, and all the rest that are handed down or intoned from the pulpits – are no more real than if they had fallen from a star.

The grounds for the diagnosis of causal autonomy are admitted to be overwhelming. It accords perfectly with both the most common-place experiences and the most abstract social theories. An exposition of the latter would be tantamount to a résumé of the principal sociological insights from the classics to the present day, ranging over and uniting such unlike minds as Comte and Adorno, Marx and

Luhmann. Émile Durkheim's 'organic solidarity'; Karl Marx's theory of the crisis of capitalism; Max Weber's theory of bureaucracy; Talcott Parsons's systems theory; Jürgen Habermas and Claus Offe's theories of late capitalism; Michel Foucault's supremacy of the institutions; and certainly also Horkheimer's and Adorno's 'Dialectic of Enlightenment', are all once again highly relevant. All agree, on very different grounds and with contrary evaluations, that the developed industrial system must be seen as an inexorable social force, which has become independent. The concept of system has been 'cleansed' by Niklas Luhmann of all intentionality, situating it in a kind of 'heaven of this world', above humanity and below the gods. This is really only the culmination of the idea with which sociology emerged from incipient nineteenth-century industrialism: the momentum of a social development founded on markets, technology, private ownership, capital investment, class contradictions, bureaucracy, and so forth. This social development is predicated on and consists of human action, but becomes independent of the latter, 'reified' into 'solidified spirit', a 'living machine', or a 'cage of servitude', to use the phrase, now on everyone's lips, of Max Weber.

Anyone who is aware of the futility of endeavour, while remaining in good voice, must present the idea on whose behalf he or she intends to argue (without reaching for the waste-paper basket, already overflowing with discarded ideas). Being cowardly and cunning, I shall find out more about the enemy's supremacy in order to fight against it. I have a notion of taking into the field the most persuasive arguments against this overwhelming evidence – the evidence itself. The causal autonomy diagnosis has the power of a truth that can only be overcome in the paradoxical form of its affirmation.

To put it another way, if any force is capable of loosening causal autonomy, of breaking its grip, then that force is causal autonomy itself. This thought becomes the more convincing where an entire epoch has let itself in for the adventure of colossal self-created hazards in order to ensure progress. It is my hunch that the successes of the institutions in our technologically advanced, highly endangered affluent society are barely distinguishable from its failures; the latter are, as it were, the other side of the coin, which shows through increasingly as the repression of hazards is perfected.

The concealed self-politicization of hazards in hazard administration

A digression on approaches: objectivism about hazards, relativism about hazards, the sociological concept of hazard

In the public arena and in specialist discussions there are above all two interpretative attitudes, poles apart as to their assessments and political consequences, competing against one another: 'natural-scientific objectivism about hazards' and 'cultural relativism about hazards'. There is much to be said for and against each of these. Current objectivism and naturalism about hazards undertaken by expensive, technological-medical diagnostic means to transform the technical formulae that gave rise to the hazards into nets for their recapture. What speaks in favour of this undertaking is the expropriation of the sense organs in the nuclear age (or, to put it the other way around, the technological monopoly on the perception of hazards and the indispensability of technological solutions). On the other hand, it is naive to take this apparently 'self-evident' fact for granted: first, because it ignores the immanent dependence of natural-scientific risk diagnoses on culture and politics; second, however, and particularly, because it is usually tied to the claim that social normality can essentially be determined by technological means. In other words, faith in technology here entails the overbearing argument that the engineers can set themselves up as judges of the rationality and irrationality of public issues and social unrest. It is a key fact of political experience that natural-scientific 'pronouncements of normality' do not assuage social protest; objectivism about hazards almost automatically evaluates and stigmatizes this as hysteria (of the media, the *Zeitgeist*, etc.).

Thus the real blind spot of technological objectivism is that it does not recognize the independent political dynamic of large-scale hazards. It supposes that a society does well to cling like eager little freshman engineers to the heels of its top technologists' arguments, and can be compared to attempting to explain parliamentary controversies from the perspective of the biological process.

Cultural relativism about hazards, for its part, can point to masses of empirical evidence in its favour: the ups and downs of maximum pollution levels, nationally and internationally, at one or various times; or large-scale hazards (nuclear power, dying forests, Aids) fiercely competing for the attention of the policy-makers. Historically, too, the cultural dependency of hazard assessments can be shown very concretely. Who is to say that the rustling of wind in the leaves, a sign for some ancient civilization of the presence of terrifying demons, did not unleash more horror than the equally imperceptible dangers of nuclear radiation?

But precisely here lies the source of the error: while correctly appraising the cultural relativity of hazard perception, there is smuggled in an assumption of equivalence between incomparable terms. Suddenly everything is dangerous and thus equally safe, judgeable only within the horizon of its social estimation. The central weakness of cultural relativism lies in its failure to apprehend the special socio-historical features of large-scale hazards in developed technological civilization. Unlike a nature animated by demons, creeping pollution is a product of social decisions and regulations that can in principle be avoided or altered. This social reflexiveness can only be neutralized socially by holding fast to fallacious principles. Demons, natural disasters, acts of God, on the contrary, remain attributable to another world, and can be passed off as the actions of divine powers; people must come to terms with the latter, but in the end only the supernatural powers themselves can be held responsible for the threat they present. Large-scale hazards of civilization, on the contrary, came into the world as opportunities, with the blessing of science and technology; they must first be 'extracted', accident by accident, from the shell of the alleged advantages by which they are protected. It is thus precisely political reflexiveness, not the level of destruction (dead, injured, etc.), which distinguishes man-made hazards from their pre-industrial 'avatars'.

These two standpoints and approaches will now be qualified and integrated into a sociological perspective: this will centre upon the institutional contradictions between the safety and control requirements imposed by the state on the one hand, and the normalization of large-scale hazards on the other. It is precisely in the context of highly developed welfare and safety bureaucracies that the legalization of decision-dependent dangers of annihilation points to the immanent social contradictions; and also to the political dynamic wherein the social subsystems (economy, science, law, politics) have become entangled in a civilization of large-scale hazards.

At least a threefold disjunction separates large-scale ecological, nuclear, chemical and genetic hazards from the (enduring) risks of primary industrialization: first, the former cannot be delimited, whether spatially, temporally or socially, and thus affect not only producers and consumers but also (in the limiting case) all other 'third parties', including those as yet unborn; second, they cannot be attributed in accordance with the rules of causality, guilt, liability; third, in so far as they cannot be compensated (because they are irreversible and global) according to the current rule of 'polluter pays', they are irremediable hazards imposed upon the alarmed safety consciousness of citizens. The calculus of risk, upon which the administration of hazards founds its rationality and safety guarantees, accordingly fails. Large-scale technological-ecological hazards have undermined the accident (at least, as a temporally and spatially circumscribed event), hitherto the foundation of the calculus of risk. The repercussions extend over countries and generations, and can themselves only be released from their individual anonymity through a statistical wrestling bout.

'Risks' are understood here (similarly in principle to the prevailing conception) to be determinable, calculable uncertainties; industrial modernity itself produced them in the form of foreseen or unforeseen secondary consequences, for which social responsibility is (or is not) taken through regulatory measures. They can be 'determined' by technical precautions, probability calculations, etc., but (and this is frequently not taken into account) also by social institutions for attribution, liability and by contingency plans. There is, accordingly, a consensus in international social-scientific literature that one should distinguish here between pre-industrial hazards, not based on techno-logical-economic decisions, and thus externalizable (onto nature, the gods), and industrial risks, products of social choice, which must be weighed against opportunities and acknowledged, dealt with, or simply foisted on individuals.

The current debate, by contrast, aims at one more crucial step (also a central concern of this book), namely, the distinction between (industrial) risks and the return of incalculable insecurities in the form of large-scale hazards of late industrialism. The latter also emerged historically out of human deeds, so they cannot be palmed off on extra-societal forces and influences; but they simultaneously undercut the social logic of risk calculation and provision. This line of argument coincides with a (somewhat crude) distinction between the epochs of 'pre-industrial cultures', 'industrial society', and 'risk society'. The typology (or more conventionally, set of hypotheses) briefly sketched out in table 1 is elucidated and developed in the following sections (cf. also my *Risk Society*, part I).

Yet not only are the hazards qualitatively new, compared with the institutions' control claim; as the welfare state has developed (see

Table 1 *Risks and hazards*

	Pre-industrial high cultures	Classical industrial society	Industrial risk society
Type and example	Hazards, natural disasters, plague	Risks, accidents (occupational, traffic)	Self-jeopardy, man-made disasters
Contingent upon decisions?	No: projectable (gods,demons)	Yes: industrial development (economy, technology, organization)	Yes: nuclear, chemical, genetic industries and political safety guarantees
Voluntary (individually avoidable)?	No: assigned, pre-existing external destiny	Yes (e.g. smoking, driving, skiing, occupation) rule-governed attribution	No: collective decision, individually unavoidable hazards; yes and no (organized non-responsibility)
Range: who affected	Countries, peoples, cultures	Regionally, temporally, socially circumscribed events and destruction	Undelimitable 'accidents'
Calculability (cause–effect, insurance against risks)	Open insecurity; politically neutral, because destined	Calculable insecurity (probability, compensation)	Politically explosive hazards, which render questionable the principles of calculation and precaution

'Liberty, equality, safety', Brückner 1972), the demands to avert and control hazards have also increased and spread to every area of life.

> If many people today, not only conservatives, consider it necessary to admit new risks into the private domain, but also into the economy and the public arena, and if there are polemics against the 'tutored helplessness' of citizens in danger of suffocating under excessive social provision, and whose personal freedom of choice is a social good to be valued at least as highly as safety, then that points to the paradox of safety: the growth of

external guarantees can clearly lead to diminishing tolerance of existing or new perils. (Evers and Nowotny 1987, p. 61)[1]

Normalization of hazards: the 'bureaucratic fall from grace'

Two contrary historic developments thus converge in Europe at the close of the twentieth century: a safety level based on the perfection of technological-bureaucratic standards and controls; and the dissemination, the imposition, of historically unique hazards which slip through the finest nets of law, technology and politics. This non-technological, sociopolitical contradiction remains concealed by the confusion of centuries so long as the old industrial patterns of rationality and control continue to hold; it breaks out as the improbable becomes probable, the exception turns into the rule. Charles Perrow (1984) calls this foreseeable incidence of discounted possibilities 'normal accidents', which are rendered more likely, appalling and shocking by the vehemence of their denial. The chain of publicly scrutinized disasters, near-disasters, and hushed-up safety flaws and scandals causes the technically oriented control claim of state-industrial authority to be shattered – quite independently, moreover, of the established measures of hazard: number of dead, danger of contamination, and so forth.

The chief social-historical, political potential of ecological, nuclear, chemical and genetic hazards lies in administrative collapse, the breakdown of scientific-technological, legal rationality and of the institutional-political safety guarantees which these hazards publicly conjure up. It lies in exposing the social production and administration of large-scale hazards for the wilderness of 'real existing anarchy' which they have become, under the conditions of their denial.

Take the case of Chernobyl, a worst-case scenario outside Germany's borders which exposed the populace to nuclear contamination; a disaster was not envisaged, let alone provided for, in the regulations or the plans. 'Existing precautions only applied to inland reactor sites and environs up to a radius of 25 km' (Czada and Drexler 1988, p. 53). This misjudgement of distances vividly illustrates the confusion of centuries upon which the administration of hazards has constructed its rationale. Thus the plans and regulations

were not applicable in law to the dangerous situation after Chernobyl, and would moreover politically have entailed the admission of a disaster situation in Germany. Accordingly, the reactions of politicians and authorities varied extremely, and seemed to depend upon chance – on whether, say, the fire brigade's Geiger counter was operational, or a professor of physics with an appropriate device was at hand. This situation was exacerbated by the absence of policy directives to the administrative sub-authorities.

Politicians and administrators were in something of a state of emergency between 29 April, the date of arrival of the radioactive cloud, and 5 May 1986, when reports from the commission for radiation protection first reached communities nationwide through official channels. The uncertainty of the situation was exacerbated by the conflicting advice and recommendations of the *Länder*, whose individual departments disagreed in their turn. The lower administrative levels reacted to the measurements and to local civic concern with a variety of initiatives, or with helpless concern . . . of the fifty-one district and town administrations questioned, 31 per cent had set up crisis committees, while others thought these were unnecessary. Where such bodies did exist, 22 per cent were composed not only of the heads of relevant departments but also of representatives from external institutions whose co-operation had been sought. (ibid., pp. 53ff)

After the Radiation Protection Act came into force on 30 December 1986, the definition of hazard was entirely turned over to the Minister for the Environment and to a federal co-ordinating office. It remains to be seen whether this move will make it possible to avert chaos.

With the emergence of the hazards for which they assume (but deny) responsibility, the institutions enter an unwinnable race against the safety claims that have been extorted from them. On the one hand, they end up under the perpetual compulsion to render still safer that which is already as safe as can be; on the other, people's expectations are thus raised, ultimately intensifying awareness to a point where even the merest suspicion of an accident makes the façades of safety claims crumble.

In this sense, large-scale hazards that reach public awareness must count as members of that rare species of event, unintentional debureaucratization. They are a kind of 'bureaucratic fall from grace', the expulsion of bureaucracy from the earthly paradise of its false claims and vaunted omnipotence. Hazards are the instrument, not yet discovered or utilized, of de- and anti-bureaucratization. They blow up the façades of (in)competence. They tear down Potemkin villages, entire city-states predicated on welfare and responsibility. They reveal infestations which, viewed from inside, had seemed proof against inspection and criticism. They unmask safety constructions built on card-houses of probabilistic assumptions. They transform cash fountains, apparently destined to confirm iron necessities, into the ruins of bad investment, memorials to technological megalomania.

The independent momentum of industrialism, in combination with the mass media and social movements, has in a sense assumed the role of the Enlightenment. Now it both constitutes and sheds light on the danger it has created. By the instrument of terror it achieves something that no book, dissident expert or social movement has been able to accomplish, however deep its social commitment. It tears up the tissue of lies, fabricated by a whole epoch, concerning the commensurability and surmountability of the dangers of self-annihilation that emerged with industrialism. Yet it cannot do this on its own: action must be taken to raise public awareness and keep it raised, in spite of the routine of industrial normalization and cover-ups.

'If our . . . comparative analysis of politics in the risk society is correct', Herbert Gottweis wrote at the end of an international survey,

> then there is a gap today between modern or modernizing societies' potential hazards and risks, always growing and increasingly subject to cultural perception and problematization, and the political handling of these technological and cultural developments. This gap articulates itself in the western industrial countries above all as a crisis of political output: against a background of growing political mobilization against risk, the ensemble of legal and administrative tools, developed for the solution of quite different problems and hence antiquated, comes up against factual and financial limits to the solution of problems . . . This gap will become ever less bridgeable by means of gesture politics or nods at the force of circumstance, e.g. energy policy requirements. The most important political consequence of a succession of environmental disasters that decimated whole populations might be a massive outbreak of civil strife and international conflicts, as can already be observed in certain regions of the Third World. (Gottweis 1988, p. 13)

The range and explosiveness of these conflicts can be gauged from the internally unresolvable contradictions in which policy and administration have become entangled, confused over the centuries. The thesis to be developed below, of the concealed self-politicization of hazards in hazard administration, is not predicated on the self-activation of a new Enlightenment, this time about technological hazards. It is not a question of declaring social movements, public protests and the attention of the mass media to be superfluous, but quite the contrary: to point them out, to clear up through argument the decisive starting point for activities (see especially chapter 7).

It was possible to mobilize the productive technological forces of society in order to transform the hazards of external nature into the ascendancy of industrial civilization. The hazards of second-order civilized nature can only be checked by an emollient, an antidote that turns the industrial dynamic against itself, i.e. by applying the logic of system to take the logic of system down a gear, through experience of the danger of annihilation, to the logic of action. Dangers of annihilation predicated on human decisions shatter the 'natural' character of civilization. With the violence of the momentum they derive from, they compel a new interrelation, a new form of interference, between the individual and society. They are one way for the supremacy of the 'industrial system' to display its own powerlessness and irrationality. They are atheism fully achieved – even in the face of 'systems theology', wherein human beings can no longer sacrifice the empowerment that is their birthright before the false gods, this time truly their own creation, of technological, economic and other forces of circumstance. They are all these things, but (as we have already said) not automatically – only in so far as the opportunity for action, which they make possible, is exploited. This is a ray of hope, the slimmest of chances amid the danger that strives to destroy all life and is capable of breaking out in many forms, apart from this late ruse of reason.

Change without change: the blindness of sociology to hazards

Post-war sociology in Germany has worried and argued about everything: whether popular consent to the capitalist welfare state should not be considered a demonstration of capitalism's effect on people; how it was possible to tolerate for so long the existence of only two complete editions of the works of Max Weber (a scandalous state of affairs, now at last overcome by the publication of the third); whether 'interpenetration' (not a sexual refinement) derives from Durkheim and means that Parsons is outmoded, or the reverse. Yet the following question has been criminally neglected by the social sciences: what does the threat of self-annihilation mean to society, its institutions, its understanding of progress and of itself; to the legal, scientific and economic system; to politics and culture? Sociology is just as 'blind to the Apocalypse' (Günther Anders) as society is – and at least in that respect, it is an empirical discipline.

Talk of the 'nuclear age', or more concretely, the question of the social reality behind the political judgement of what (in political language) is 'technically feasible, but politically unrealizable' has remained a closed book to sociology. This is the more surprising because in spite of the inevitable disagreement over what, why, in what respect and for whom something constitutes a 'risk', there would probably soon be unanimous agreement on one basic historical fact: namely, that the second half of the twentieth century has distinguished itself – by virtue of the interplay of progress with the possibility of annihilation by the ecological, nuclear, chemical and genetic hazards we impose on ourselves – not only from the first phase of industrialism, but also from all the infinitely various cultures and epochs in the history of humankind.

The more this issue concerns the public, the more puzzling becomes the reluctance of social theory to engage with it. Whatever reasons may be propounded for this fact – the tranquillizing effect of *post-histoire*; obliviousness to the social-historical dimension of the present and the future; servile dependence on the authors of the classics (who really could not have known of this); the piling-up of details, which the methods and solemnity of the *arriviste* science of sociology would appear to enjoin, etc. – these do not really matter. What is important is the result: sociologists (with few exceptions) assiduously argue and carry out research assuming that the safety problems in the nuclear, chemical and genetic industries at the end of the twentieth century can in principle be overcome with the same social institutions as those of pin factories in the early industrial era.

Not the least of obstacles is the well-worn divide between technology and society: the danger gives itself out to be physical, ecological and medical. It stems from technological achievements and can only be dealt with, formulated, defined, minimized, maximized, by technological means. Yet the natural-scientific formulae that trouble the world are only a technologically mystified mode of society's encounter with itself in its history, in its own decisions, standards of acceptability, power relations, conceptions of justice and foundations of rationality. Implicit in these are conventions based on politics and cultural values and perceptions, the legal judgements that have realized current law, and an understanding of technology and science that derives from 'occidental culture' (Max Weber).

Thus it is not the new technologism *per se* of our disposition over life that characterizes the new epoch. Nor is the mere threat of total annihilation without social precedent. The Apocalypse is a trauma of humankind, present in every culture, religion and sacred text. Yet on the basis of what privileged insight is anyone to say which is the more probable: the annihilation of Europe by a (previously discounted) accident, or by the righteous anger of one god or another? One difficulty here is that one is barely distinguishable analytically from the other, especially in retrospect. The possibility of the annihilation or creation of life is based on the social institutions by which it is administered, encouraged, discounted and perfected – in contradiction, moreover, to the relevant institutional norms, historical foundations and cultural expectations to which these social institutions attach. This interplay (different from one country to another) between annihilation and progress, justice and intimidation, wealth and destruction, and rationality which promises to defuse hazards but endangers everybody, gives the hazards their cultural, social and political character.

Physical annihilation itself remains an unreal scenario, because the Apocalypse simply cannot be experienced. Yet the representation, the picture that is painted, the perception of decision-dependent self-annihilation in the developed security state – that is a unique political reality, whose deep effect can only be guessed after Harrisburg, Chernobyl, Bhopal, etc.

Arnold Gehlen was speaking for many others when he declared in 1963, casually and with relief, that the Enlightenment is dead, only its consequences continue to operate. In today's developed industrialism of hazards I should like to respond, with hope and a sigh: industrialism's premises are dead, the Enlightenment is beginning anew.

Perhaps the hazards and all the terrors they spread are a kind of extra tutoring in the historicity of industrialism. Perhaps, against all the prevailing assumptions of the end of history and the self-sufficiency of industrialism, they will form a bridge: from the industrial stone age of the past to an enlightened, future industrialism of actions where the basic questions of 'progress' are extricated from the anonymity of organized non-responsibility, and new institutions of attribution, responsibility and participation are created.

Critique of bureaucracy: hazard administration as a hazard to the administration

The signs seem overwhelming that political and administrative control claims are sustainable even in the civilization of large-scale hazards: the 'symbolic detoxification' policy, which enjoys great popularity across the trenches between the political parties, leads directly to the staging and perfecting of a cosmetic treatment of risks; and the latter is now greatly facilitated in the non-empirical space of theoretical constructs. If Chernobyl gave an example of a failed information policy, then I dread to think how a successful one would manage risks. What does 'hazardous' mean? Under conditions of increasing normal pollution, the definition of hazardousness is a question of official standards and of how they are established. Where all else seems to elude political intervention, and contamination rises by leaps and bounds, this question becomes the central lever to apply to the failure of a policy of definitions to overcome the increasing hazards. Abstraction, dependence on expert opinion, legal and scientific power over the definition of terms: these would appear to turn the citizens and their anxieties into the football of hazard administration – were it not for the counter-force of hazard itself, under the spotlight of the mass media.

Chernobyl uncovered the real existing administrative chaos, smashing the myth of 'residual risk' on the anvil of its reality; and it disclosed the insecurity of the 'security state', the lack of prior provision. If a fire breaks out, the fire brigade will arrive; in the event of a traffic accident, the insurance company will pay. This interplay of earlier and later, of security in the present moment because precautions have been taken even for the worst imaginable case, is annulled in the nuclear, chemical and genetic age. The shining achievement of nuclear power plants is not only to have made redundant the principle of assurance in the economic sense, but also in its medical, psychological, cultural and religious senses. 'Residual risk society' is a society without assurance, whose insurance cover paradoxically diminishes in proportion to the scale of the hazard.

No real institution, and surely no imaginary one, would be prepared for the worst-case scenario that is threatened; and no social order in this extreme case could guarantee the cultural and political constitution. On the contrary, many of them specialize in the only available recourse, which is to deny the hazards. For remedial action, which provides a guarantee of safety even under dangerous conditions, is replaced by the dogma of infallibility that the next accident refutes. Science, the queen of errors, turns keeper of the taboo. Only 'communist reactors' but not German ones are the empirical products of human hands, able to throw every theory onto the scrap heap. Even the simple question 'but what if . . . ?' is lost in the vacuum of remedial measures. Political stability in risk societies is accordingly the stability of thoughtlessness.

The harsh glare of the burning Sandoz warehouse and the Rhine disaster shed light not only upon the hazardousness of the chemical industry, the calculated ignorance of managers and authorities and the lack of state controls, but also on the fundamental difficulties of the judiciary in prosecuting chemical crimes. Even where substances of demonstrably and culpably high toxicity have flowed into the Rhine, it remains unclear whether anyone can be made responsible. The criminal law of Germany cannot be applied to a company as a 'legal person'. It is only aimed at 'natural persons', and even if the latter were found, their commission of a 'culpable action' would have to be demonstrated in the social maze of company law procedure. And in the end, the *Atomgesetz* (Nuclear Act) for example 'threatens' to impose a fine on conscious offenders (see the detailed argument in chapter 6).

To take up one last case previously deemed impossible, the bribery scandal involving the atomic energy companies Nukem and Transnuklear revealed that incredible sloppiness in handling radioactive materials of all kinds was the rule in these centres of Germany's nuclear fuel network, that plutonium, the raw material for nuclear bombs, travelled freely across European borders, and that waste circulated illegally in the nuclear power industry, which is purportedly so security-conscious. Under the spotlight of the mass media, the nuclear power industry thereby achieved what years of unavailing criticism of nuclear power had been unable to: the refutation of the safety and control claims of an entire industry and its supporters in government and administration. The inability of pro-nuclear politicians to dismiss, out of hand, suspicions of secret deals with Libya or Pakistan thus made it clear that even government policy has to sell out on its safety claims. The mere suspicions that one entertains are ultimately enough to make the façades crumble.

Large-scale hazards: a scandal covered up

In the highly developed security state with its citizens' awareness of safety issues, large-scale technological hazards insusceptible to compensation constitute an institutionalized scandal which can only be covered up, but not eradicated, by the legal and scientific constructs of non-hazardousness. Hazards which come into existence with the blessing of technological and state authority place authorities and policy-makers under the permanent compulsion to assert that these hazards do not exist, and to defend themselves against penetrating questions and evidence to the contrary. The result is that politics is identified with the safety façades, and its foundations begin to crumble along with them.

It seems no exaggeration to say that the hazards of annihilation are the motor of bureaucratic expansion and of the immanent self-refutation of its rationality. Max Weber proposed that the deputy sheriff of the 'charismatic leader' should keep bureaucracy within bounds; the implementation of this counter-bureaucratization, become independent, has been undertaken by the bureaucracy itself in its handling of large-scale hazards.

Even where technology is constantly being improved, and the probability of, e.g., a reactor incident is continually being minimized, that probability can only ever be diminished; a worst-case scenario cannot effectively be excluded. The converse of this, however, is that the threat to all is legalized. Probabilistic safety is compatible with accidents. Probabilistic safety and the annihilation of all life: this is mathematically, but not socially, feasible. What jeopardizes everyone, no matter how minimal its probability, slips through all the nets of law, science, technology, medicine, politics. The contrary logic, upon which the developed bureaucracy is based, runs: it jeopardizes every-thing, so it does not exist. We cannot exclude or master it, so it is legal. Whole armies of bureaucrats busy themselves in splitting hairs: what jeopardizes everyone is first minimized, then legalized, then attributed to fate and imposed against popular protest upon the population.

Whether they like it or not, citizens are delivered up to the risk of nuclear energy, chemicals and genetic engineering. Even a country that decides to forswear nuclear energy production is affected by nuclear plant failures in neighbouring countries. Even if every state were to decide today to switch off its nuclear power plants, many generations would have to guarantee that the radioactive potential had been securely sealed in. All of that makes it apparent that the degree of danger and the administrative reality of the state's safety guarantees belong to different centuries. The character of the hazard rescinds the claim to technological control, in the perfecting of the latter. This gives rise to an objective ambiguity wherein assessments vary hugely, and the political barometer of mood can run haywire: periods of extreme unrest are followed by periods of aggressive placidity. Both reactions have the same cause: the unimaginability of a 'legitimate' hazard with which everybody has to live.

The contradiction between safety and control claims and the nor-mality of disasters in large-scale technological systems leads, as Charles Perrow establishes in his comparative international study of disastrous accidents, to completely standardized modes of cover-up and reaction among those responsible in the economy and in politics.

Representatives of government authorities regularly invoke the safety of nuclear power plants. Once news has reached the outside world, the accident and its possible consequences are trivialized as far as possible, and once an incident can no longer be denied, the universal, standard explanation is always either 'human error' or 'operational fault'. 'Operational faults' are corrigible, while faulty systems can only be completely reconceived or discarded (Perrow 1984, p. 7). Finally, the systemic contradiction of hazard normalization manifests itself in the need for all ensuing investigations to fight the deliberate suppression of hazards.

Hazards of the nuclear and chemical age are socially, as well as physically, explosive. The centre of this social explosiveness is located in the administration, in public opinion, the economy, politics, but not necessarily at the site of the occurrence, accident or scandal. It is not at all a physical process, and accordingly cannot be described or measured in units of radioactivity, rems, becquerels, etc. Liabilities, legal claims, the principles of calculations, legitimations 'explode' socially upon contact with reality. Admission of the danger coincides with the admission that everybody who has been right until now, including all the institutions, was mistaken and has therefore failed. The obverse of the presence of hazards is the failure of the institutions, which derive their justification from the non-existence of hazards. Therefore the 'social birth' of a hazard is an event as improbable as it is dramatic, traumatic and socially convulsive.

Precisely because of their explosiveness in the social and political domains, the actual character of hazards is that of grotesques: ambiguous, open to interpretation, akin to modern mythical beasts which appear now as a worm, now as a dragon, depending on one's point of view and interests. Thus one need only keep in mind the consequences of the acknowledgement of hazards in order to comprehend the power behind the interpretative antitheses, and their insurmountability. The ambiguity of risks is also a function of the political earthquakes and upheavals that official clarity about risks would be bound to set off. The extent to which contradictions nonetheless exist between public opinion and official policy indicates how far the unquestioned assumptions of industry have already been eroded by the acid rain of hazards.

One early phase of this erosion of the legitimacy of bureaucratic supremacy can be depicted in terms of the antitheses between the internal and external constructions of hazard and of normality. The cultural experiences and expectations of a society sensitized to hazards collide, in the economy, in science and in the legal system, with the routines of the industrial-bureaucratic risk calculation and legaliz-ation of risks (chapters 5 and 6). This difference of world view between, say, the authorities and policy-makers on the one hand, and social movements and the mass-media public on the other, intensifies to the degree in which the bureaucratic principles of rationality and justice, as applied to hazard normalization, are themselves submitted to critical scrutiny; that is, the calculus of risk and the rules for assigning responsibility and granting compensation are convicted of their historical inadequacy. Then the state tools for normalization fail, and the contradictions between institutionalized safety guarantee pledges and hazard legalization break out into the open.

Without diminishing the decisive role played by social movements in the public disclosure of hazards – that is, without stirring up the chicken-and-egg question – one can say that the stability of the social critique of hazards is predicated not only on the stability of social movements, but also on the longevity, public acknowledgement and technical insurmountability of the hazards which the institutions of industrial society have let themselves in for. The hazard of annihil-ation, a result of decisions taken and thus in principle avoidable, is one of the political forces. In so far as it is socially perceived, the violence of its threat to life causes the institutional safety pledges to unmask themselves, as it were. The hazard of annihilation triggers off a chronic devaluation of political and technological rationality claims – at first through accidents which flatly contradict declared safety guarantees, under the eyes of the world media. This provokes insight into the fallibility of all institutional guarantees, until finally it becomes imperative to prevent the threat constituted by hazards. Nothing happens, yet a great deal changes: the system of production and its 'second adjutants', consisting of politics, science and the law, are now thoroughly scrutinized for potential hazards and conceal-ment practices. Not technology but technocracy ends up in the dock. Henceforth, not only hazards and pollution are complained of, but also that which most palpably intervenes in life, eluding consent and dissent.

The cultural dependency of the calculus of risk: the dilemma of the technically-oriented administration of hazards

Max Weber clearly perceived the compulsive character of bureau-cratic development: its malleability as an instrument of 'rational domination'; the calculability of its decisions and implementations, internally and externally, which is also the ground for its 'elective affinity' with the capitalist organization of industry, its legalism and specialization, guaranteeing control and competence. All this provides the foundation of and acceleration for the rationalization process that modernity has put into gear, and that is substantially a process for the realization of a bureaucratic world order. Aside from this paean to 'bureaucratic rationality', Weber also invoked the tragedy of this development (especially in his late writings). In the end, he found recourse from the cage of bureaucratic dependency only in the authoritarian counter-force of a messianically gifted leader.[2]

We children of 'post-Nazi' Germany can hardly laugh at that. But even laughter would still not eradicate the problem (and the helplessness of this 'solution' gives an overwhelming indication of its gravity). Hannah Arendt, too (1970) saw the drift towards rubber-stamped radical and irresistible changes as a principal factor of the worldwide revolution. The impossibility of assigning responsibility and identifying opponents produces generalizations void of particulars, and thus meaningless. Ralf Dahrendorf (1980) asks whether the escapism of our own day, from massive drug consumption to pointless terrorism, is not a set of variations on this flight from the ineluctability of 'bureaucratic rationality'.

On the one side, there is a surfeit of normative critiques of bureaucracy, virtually an all-party anti-bureaucratic consensus (which usually 'skewers' the other side of its logic and performance): abstraction from people and historical circumstances; the 'inwardness' of decisions, which can be perfected to the point of an incorrigible ahistoricism; the separation of organizational ends from the motives of the agents, who turn bureaucratic organizations into compliant tools. These arguments have become almost as autonomous as the bureaucracy, thereby accompanying its victory parade as the shadow accompanies the wanderer. In contrast, hardly any starting points for a systematic counter-politics have become visible until now, with the exception of Michel Crozier's 'bureaucratic vicious circle'. That means, however, that the characteristic logic of bureaucracy compels agreement with it. There is great pressure on any sociologist or sociology to conform in declaring that bureaucratic supremacy is, if not good, then (to put it more sharply and less contestably) rational.

Thus perceived, there might be a confusion, in Weber's theory of bureaucracy, between supremacy and rationality; this, though it seems to be grounded in the special bureaucratic claim to calculability and efficiency, is quite fatal for the proponent of a 'value-free' analysis who does not want to expose himself to the suspicion of being simply pro-bureaucratic. Under the spell of a tragic fatalism that holds Weber in its thrall, the question Weber also addresses, of the historical contradictions between the 'formal' and 'substantive' rationality of bureaucratic rule, remains sententious, a product of the ethic of conviction, and fundamentally unrealistic. To that extent, it is alien to Weber's perspective to suppose that the cage of bureaucratic servility might be a house of cards, and that large-scale hazards can effectively make it tremble.

The cultural dependency of the calculus of risks, the necessity of distinguishing between destruction and acceptance, the impossibility of compelling acceptance or even protest through the technological minimization or dramatization of hazards (cf. chapter 2 above) – all this strikes at the heart of the hazard administration's technology-centred conception of control and safety.

Where consensus about progress begins to crumble, 'rationality of risks' becomes both necessary and impossible. It becomes necessary in order to manufacture acceptance, and impossible because its rationality shatters in the spectrum of cultural perceptions. On the one hand, in the civilization of large-scale hazards, the administration is under pressure constantly to submit its control and safety claims to proof. On the other hand, its means of proof fail; indeed, they are devalued, convicted of their systematic inadequacy, in precisely this process of cultural pluralization of risk perception. To the compulsion to present a hazard-conscious public with sedative calculations there corresponds the growing insight that the technological objectivity of risks, upon which the administration's safety pledge is founded, is borrowed from culture. Neither experiments nor mathematical models can 'prove' what human beings are to accept, nor can risk calculations anyway be formulated solely in technological-bureaucratic terms, for they presuppose the cultural acceptance they are supposed to manufacture.

Hazards are subject to historico-cultural perceptions and assessments which vary from country to country, from group to group, from one period to another. In France, visitors stream through nuclear reactors on public holidays, in order to view and admire these monuments to technological creativity. In the cities of China, maximum pollution levels that cause us to protest are exceeded by many times, without thus far having destroyed normality. Aids is (obviously) considered less of a risk by the old than by the young, who have to adapt their sexual behaviour to it. And each of the scenarios that derive from the big risks – nuclear power, ecology, chemicals and

Aids – has a different political affinity, posits a different change as imperative, allows for a different construal of the blame; these appear 'realistic' or 'dramatized', depending on political ideologies and standpoints. Accordingly, one can observe in society's confrontation with these issues a political struggle between the themes for competing big-risk scenarios, each with its own politically opportune consequences.

All of this serves to qualify the purely technological calculation and containment of risks, since the calculations are now no longer thought of as arbitrators but as protagonists in the confrontation, which is enacted in terms of percentages, experimental results, projections, etc. Risks are social constructions disposing over technological representations and norms. An acceptable risk is, in the last analysis, an accepted risk. In the process, what appears unacceptable today may be routine tomorrow, while previously quotidian practices suddenly fill one with anxiety and terror in the light of new data.[3]

Purposive rationality and rationality of risks

There used to be the progressive faction on one side and the critique of culture on the other. This stable world of friends and enemies of progress can be traced from the Rousseau/Voltaire debate right down to the cultural criticism applied by the early Frankfurt School to mass culture and 'bourgeois science'. These oppositions no longer obtain. Established risk research has assimilated the critique of culture. Ecological critique disposes of all the conceptual tools of science. Frequently, too, there is a scientific premise lurking in even the most committed people's consciousness, turned rigid with despair: one despairs because one believes in a specific doomsday model, not realizing that it is disputable. Doomsday visions are also covert, anti-scientific superstitions about science. In other words, the co-ordinates have not only shifted, but mingled; and, along with them, the rights to assess and 'rationally' control the threats.

Weber saw the alternatives for the development of modernity in a 'value rationality', which lives in prevalent traditions and whose 'logic' remains integrated into them, and a 'purposive rationality', which determines the means deployed and unfolds their technological rationale. In the modern age, according to Weber, purposive rationality clearly comes to predominate, perpetually narrowing down the irreconcilable battle of creeds. This has happened. Yet, in its realization, the whole frame of reference opposing purposive and value rationality has shifted, been cancelled. The purposive rationality that has been implemented, demanding today that valuations be dressed up in numerical form, has, as it were, absorbed value rationality; thereby, however, it has sacrificed its own certainty and mode of appearance, and undergone a social metamorphosis into a 'rationality of risks and hazards'. Hazards still seem to be calculable in accordance with the old pattern of purposive rationality, indeed they become so with the deployment of statistics and research capabilities. Yet anyone who wanted to reduce purposive rationality to the point of absurdity could surely find no more appropriate method than the purposive-rational self-refutation of the calculus of risks.

Purposive rationality owes its superiority precisely to the shedding of evaluative perspectives, and to the strictly immanent jurisdiction of the rationality of means for any purpose whatever. This turns into its contrary in the risk calculation. As risks become politicized, everybody has to pick a number out of the hat, and thus hazards rob public morality of its last bastions. At the same time, however, no one any longer has privileged access to the uniquely correct calculation; for risks are pregnant with interests, and accordingly the ways of calculating them multiply like rabbits. Purposive rationality becomes overextended, insecure and value-dependent at once; its realization turns into a theatre of the absurd, because everyone produces ever more contradictory results, with ever more meticulous methods.

That leads to the wrangling over standpoints, calculation procedures and results, in which legal, cultural and scientific standards enter more and more openly into conflict with each other. One, no less unintended and radical, 'side-effect' of this is that the principles underlying the whole calculation are now hardly worth the paper they are written on; and thus everyone has to revert secretly to sound common sense, in one way or another, if they are to remain capable of acting at all.

There are many sources in risk calculation for this uncertainty. Purposive rationality presupposes the reference point that is at issue – the purpose itself. Risk rationality has an open, indeterminate horizon. One must think around the corner of the future, as it were, for the unseen or neglected to become crucially important. It is for precisely this reason that the most various consequences can be predicted; they come back with a monstrous babble of contradictory advice. How is one to quantify medical repercussions (it remains an open question whether they are articulated in terms of kidneys, lungs, allergies, etc., but these parameters must also be decided upon and weighed against one another) along with economic ones (quite aside from the various 'solar systems' of economic models)? Should one multiply, take the sum of the squares, or the reverse?

Purposive rationality has also based its victory parade on specialization and differentiation. Hazard situations and risk rationality turn this logic, too, into its contrary. Certainly, one can continue to subject parts of the whole to specialist scrutiny; to examine causes and to place them at the disposal of these interests; to dissect causal chains. But that alone does not help us any further. The risk puzzle can only develop some sense of utility within a context that unites fields of specialization.

From one side there come statistics, from the other, cultural acceptance. Each side tries to make itself independent of the other, attempting to forge ahead along the old paths, in accordance with its characteristic logic. Yet each must make use of the other. Without cultural standards, all calculation remains empty; without science and experimental results, any cultural stance is adopted blindly. Either may instrumentalize the other, ignore it or utter the magic word 'irrationality' to banish it from the circle of recognized interlocutors, but precisely this proves their reciprocal dependence.

Yet unfortunately the converse also holds. Without legal norms, e.g. legislated maximum safe levels, scientific experiments for estimating risks lose all their sense. What one side must presuppose, the other derives from some place where it can never arise. In this sense, politics and administration are perpetually waiting for experimental findings, in order to derive from risk research the maximum safe levels – which must be prescribed to risk researchers. These experiments, in their turn, represent only an establishmentarian attempt to persuade, by force of mathematical weapons, a public culture turning sceptical. Cultural acceptance, which is supposed to leap out of experiments, must in reality be fed into them if the experiments are not to be empty of significance. Experiments are carried out on mice, rats, etc., which do not need to be persuaded that they are being poisoned. Yet it is only by a leap into the infinite, from which nobody has yet returned, that one can draw conclusions fron rat experiments as to the effects on human beings. Thus rats have ultimately not been persuaded but killed, in order to convince humans; yet human beings, who can never be persuaded in this way when in full command of their senses, can, like rats, have their health and lives jeopardized.

Maximum acceptable levels, however, are no mere castles in the air, made out of experiments, interpretations, policies and cultural acceptance. The destinies of markets, and hence of companies, industries and jobs, depend on them (although it is not all that easy to perceive just how). Thus all the subsystems, which have become independent, are interconnected 'behind' the system via the constructs of maximum acceptable levels. In any case, they are so in the legacy of irresponsibility that is produced. This complexity of dual-level irresponsibilities constructed out of responsibilities may scarcely be provable, if at all, by sociological analysis, because the latter must always employ an external viewpoint, making use of the regulations, the spoken word and hence what is justifiable. Its reality makes itself manifest in its effects. Though it is certainly wrong to blame a television programme about worm-infested fish for plunging an entire industry into misery, that is precisely its effect, no matter what network of unperceived interconnections one posits. Thus, with the socialization of risk, growing hazards lead to an explosive mixture of responsible irresponsibility, in which possibly tiny sparks of information can endanger whole markets that play no part, or only a very small one, in the production of hazards (cf. chapter 6).

On one side there are interpretations, numbers, norms, axioms, i.e. things made of mind and paper, which may rapidly look very different if viewed the other way around, as rock-hard constructs of civilization: money, power, markets, dams constructed against floods of claims, points for social identities and protests to crystallize around, pillars of political responsibility and safety claims, which, conversely, are perpetually weakened by the science they are supposed to uphold. Maximum acceptable levels are a kind of mother at the centre of an infernal machine, become independent, who continually intends to produce security, thereby increasing insecurity.

Thus normality's victory over the anomalies to which it gives refuge, becomes a Pyrrhic victory. The danger grows as it is suppressed, and produces collective experiences that become increasingly antithetical to established risk calculations. The identification of risks with technological risks, which is supposed to secure rationality against the pluralism immanent to it, contradicts the experience of the population, which is troubled in particular by social, psychological and ecological repercussions. Risk rationality's motto, that any technologically conceivable option can be implemented, now means that a technological impossibility must be foisted on the general public as something inescapable and normal. The greater the difference between established, technologically based safety claims, the greater the contradiction as collectively experienced between risk calculations and real hazards: the brilliant steel walls of competence, put up by the technology-centred risk administration in order to contain hazards, collapse and expose to view the bureaucracy, which has now become politically malleable.

The conditions rehearse the uprising: annihilation hazards as an independent revolution

For at least one historic moment after Chernobyl, the revolutionary power of the nuclear peril was out in the open: in a poster campaign, the Greens urged the people to eat only tinned vegetables. The seal of approval that designates free-range eggs suddenly came to represent a warning signal, comparable to a death's-head emblem. Products from hen-flaying battery farms were in demand. The 'muesli-freaks' queued up for dried milk and frozen foods. Wherever national sovereignty had been partially devolved to the German *Länder*, petty states were suddenly in control of maximum pollution levels. Depending on whether one lived in the Rhineland-Palatinate, in the Saarland or in Hesse, the world was either in good shape or totally poisoned. Nothing contributed more to the general confusion than the routinely repeated assurance that there was 'no hazard to health' – while it was simultaneously recommended that one should keep one's children well away from the sandpit, walk on the lawn only in case of an emergency, and remove one's clothes and shoes immediately after rain; while it went without saying that children and dogs were to be showered every time they returned from outdoors. Experts were always on hand to deduce all the advice given and measures taken by the authorities, no matter how contradictory, from scientific findings.

These and other punchlines of the real-life nuclear satire added up to far more than a change of priorities from fresh vegetables to canned, bland products. An event takes place: coming from another world, centred upon a point 2,000 km distant, it completely eludes everyday modes of perception, and remains a pure media event; yet it renders the rules of everyday life obsolete, in a way never achieved by any parliamentary upheavals, mass demonstrations or social movements in Germany. Façades of responsibility, of technical competence, of the ability to master a technology and calculate its consequences, and of prior provision in the event of a catastrophe – façades erected and maintained for years by the political and industrial elites, and seemingly secured until the end of time by the investment of billions – collapsed under the assault of contradictory data.

One can say that Chernobyl was a kind of objective demonstration by the conditions against themselves. The global experiment of nuclear energy has meanwhile taken over the role of its critics – perhaps even more convincingly and effectively than the anti-nuclear movement itself, even at its apogee, could ever have done. This is not only made apparent by the worldwide, free negative advertising, at peak viewing hours on the news programmes, and on the front pages of the newspapers (compared to which even full-page advertisements by the pro-nuclear lobby, designed to assuage people's fears, now resemble the huddles of high-minded demonstrators inside the nuclear plants' reinforced perimeter fences). It is also made apparent by the fact that the whole area from the North Sea islands to the Alpine refuges has learnt overnight the language of nuclear power's critics. Under the dictate of necessity, humans have graduated from a kind of crash course on the contradictions of hazard management in risk society: on the capriciousness of maximum safe levels, the arbitrariness of calculated hazards, the unimaginability of long-term repercussions and the possibilities of anonymizing them statistically, etc., a course that is more comprehensive, clear and vivid than any instruction the most critical critique might have given or foisted on them.

Yet the nuclear hazard is by no means a revolutionary force in the customary sense of the word, but rather a revolution as the inversion of revolution, inverted as in a mirror; a revolution without revolution, without a subject. It is a revolt instigated and sustained by the conditions against themselves; the latter, having normalized the hazards, are now carrying out their denormalization and destabilization, with their self-depreciating strategies of talking them down.

Various forms of revolution have been contrasted to one another: *coup d'état*, class struggle, civil disobedience, etc. What is common to all is their empowerment and disempowerment of social subjects, and their concentration on political conditions. Revolution as process, become independent, as a concealed, latent, permanent situation wherein the conditions get caught up against themselves, while political structures, property relations and power relations nonetheless remain constant – this possibility has not to my knowledge been considered, let alone thought through. Yet the social force of hazard fits precisely into this conceptual schema. It is a product of the deed, requiring no empowerment or licence; it imposes itself, while dressed in the garb that allows progress a free passage through all the controls: science, increased productivity, lighter work, higher employment. Once they have come into existence, their coming to consciousness stands in contradiction to all the promises that stood godfather to their birth.

The principal opponents of nuclear energy (and of the chemical industry, etc.) are not the demonstrators at the perimeter fences or the critical public, in spite of their importance and indispensability; the most convincing and enduring opponent of the nuclear power industry is – the nuclear power industry itself.

The counter-force of hazard

Nearly everything rebels against this interpretation: the return, a few weeks after Chernobyl, to the calm prior to it; the flexible tiers of authority, which, in battles over definitions, successfully absorbed the waves of protest, before turning them against themselves; the 'technologizing away' of hazards, from boiling them down in risk calculations, through the exaggerated superlatives of safety technology, to wishing them away in calmative medical formulae concerning the harmlessness of low-level contamination (of which it is simultaneously admitted that little is known). The official report, adroitly linking an admission of failure to the promise of improvements, concluded that the whole thing had ultimately been an informational disaster. With the standardization of maximum safe levels, the centralization of information systems, and professional public relations exercises, worst-case nuclear disasters (which are out of the question, anyway) could in future be weathered without any trouble.

One might call this the law of inverse significance of hazard and institutional perception: the greater the range, obviousness and uncontrollability of hazards, the less significant they become in official representations of safety. Official institutions use all the instruments at their disposal (the judgements of experts, maximum safe levels) in order to claim that the hazards are harmless, and hide them in state monopolies over definitions. By the institutions' lights, the whole thing always turns out to be 'much ado about nothing', a 'nothing', though, that must always be inverted, from its dawning hugeness into its nothingness, through a multilingual self-refutation across the airwaves and on television screens from Bavaria to Siberia. With that, however, the construction and consolidation of that natural-scientific benignity, in which all the accidents end up, betrays the extent of the immanent jeopardization of the system, which is precisely what is supposed to be hidden and averted. Is a 'nothing' to convulse the firmly established social systems of high industrialism more than all the 'somethings' that claim so much of its attention? Or does the actual proof that a 'nothing' is nothing consist in its robbing a highly industrialized world's systems of independent momentum?

On one side, one can pile up all the plausible arguments for the institutional non-existence of self-annihilation hazards; indeed, one need not deprive the institutions that reign supreme of a single ray of hope, but can moreover take into consideration the dispersion of social movements and the limitations of their political effectiveness, in order to recognize no less realistically that all this is also checked by the objective counter-force of hazards: it is constant, enduring, not tied to the interpretations that deny it, present even where the demonstrations have long since weakened. The probability of improbable accidents increases with time, and with the number of big technologies that are implemented; each 'event' awakens memories of all the others all over the world.

Yet this tendency of hazards to unmask themselves remains tied to specific conditions: (a) the social legitimacy and centrality of the values jeopardized, (b) the concealed self-contradiction of the institutions, (c) conflicts of interest in the economic camp between those who profit and those who lose by risk, as well as (d) the (comparative) independence of the press. Put differently, the ecological movement's political options are not only, perhaps not even, primarily grounded in the ecology movement itself, and to that extent they cannot be adequately understood in terms of organizational problems, ideological vagueness, etc. The 'impact' of the political environmental groups is far more an expression of the scope, central significance and scarcely concealed latency of systemic contradictions in the handling of legalized hazards, raising them into public awareness. Where that succeeds, it becomes clear that the guarantees of safety, order and affluence have also manoeuvred themselves into the twilight of the opposing role.

The environmentalist critique is biting, corrosive, because it wants to preserve the things that conservatives in every camp put at stake, in open self-contradiction, with reckless industrialization policies. Years ago, Erhard Eppler used the term 'value conservatism' for this partiality on the part of critics of uninhibited industrialization for that which is to be preserved. Thus whoever argues and demonstrates against legalized 'normal' poisoning need not fight, as if in a second French Revolution, for the victory of principles that are supposed to change the world. He or she need 'only' make the accusation that is on everybody's lips – that the claim purely and simply coincides with the reality of its contrary. It is precisely this 'socially scandalous' aspect of hazards which entices the mass media with the promise of high circulation levels and viewing figures, and leads journalists to ask trenchant questions and to produce 'headline news'. Where cynicism reigns, corpses bring the news to life, and the disasters of the age are top priority. Ecological journalism does not thereby defend any partisan interests, but only the common good; and moreover, at its most sensitive spot, the highly legitimated values of health and life in a medically fit, secular society with its endless succession of market-led fitness and nutritional crazes. Herein lies the reason for the attentiveness of the mass media, and not in the media *per se*.

The uncovering of hazards, moreover, does not affect the total economy (as wage overheads have done from the beginning, and continue to do in the conflict between labour and capital) but individual industries, enterprises, products – others, on the contrary, profit from them. So there are some that cry out in pain at destroyed investments, while others extract competitive advantages and capital from risk definitions. Often enough, it is only a question of time – and of the companies' market positions – before any of them shifts from one camp to the other (see chapter 6).

Precisely this gives rise to the objectified power of the critique of hazards: it enables central social values to be sustained against the organized irresponsibility of the security state's pseudo-guarantees. The universality of hazard elevates the critique into a spokesperson for everybody. The critique also opens up markets and opportunities for economic expansion. All this gives the mass media a leading role in sounding the social alarm – so long as they dispose of the institutionally guaranteed right to select their own topics. What eludes sensory perception becomes socially available to 'experience' in media pictures and reports. Pictures of tree skeletons, worm-infested fish, dead seals (whose living images have been engraved on human hearts) condense and concretize what is otherwise ungraspable in everyday life.

The automatism of progress, on the one hand, and its critics' conformity to values and to the market, on the other, are certainly opposed dynamics of unequal weight, which clash within the framework of western mass democracies. Where social movements have sounded the alarms, their critiques can establish themselves, to a degree. Holding politicians to their promises of safety; reorienting the economy towards the risk-sensitized, high-income consumer of the future; the development of new global markets; the tabloid press's concern to maximize circulation, etc.: these provide the public critique of hazards with long-term executive powers. It can lay claim to more than the mere perception of justice, standing in for a judiciary that is obliged by its very principles to tolerate technology: it even apprehends the interests, in the beautiful terminology of orthodox Marxism, of 'total capital' against those of 'purblind individual capital'. For if the growing wretchedness of the wage labourer contradicts the long-term interests of capital exploitation, how much more so does the poisoning of consumers.

The world has become the test site for risky technologies, i.e. also a potential refutation of the safety guarantees of state, economic and technological authority. Not only the pollution but also the administrative and political repercussions of major disasters have become boundless. Here, authority hangs by the perhaps even finer thread of foreign company decisions and state authorities. The weakest link in the global chain of hazards also decides the future of highly developed industries.

The system of institutionally heightened expectations forms the social background in front of which – under the close scrutiny of the mass media and the murmurs of the tensely attentive public – the institutions of industrial society present the dance of the veiling of hazards. The hazards, which are not merely projected onto the world stage, but really threaten, are illuminated under the mass-media spotlight. The public wants to be entertained; it pays, and delights in the thrills and spills of the real-life technological thriller, in which theatre and reality have changed places worldwide. In the grammar of entertainment, the superlative of the thriller or video is the daily news. Here one discovers how ridiculous the dark fantasies of the film-makers are, compared to reality. The invisibility of the hazards produces the cultural blindness that finally robs TV reality of its suspension of disbelief, and invests it with actual reality. As the didactic drama of Winscale shows, if mass-media access to hazards is not officially prevented by the imperceptibility of nuclear contamination, then every hazard, every accident, every suspicion becomes a nationwide didactic drama in the critique of technology and progress. It is performed by technology and industry themselves, which, as self-financing members of their critics' inner sanctum, are most competent to represent and concretize the refutation of technocratic rationality. They do not have to play themselves, only to act as if they were who they are.

The piece cannot be dropped from reality's programme because the hazards themselves are there to stay, and running the show; it takes on the form of a real-life satire, endangering the actors' jobs: there are perpetual declarations of security which thereby foster insecurity. The dance of the veiling of hazards is presented with involuntary, and therefore consummate, sick humour as a dance of unveiling, in endless variations.

The tools used in the presentation are the arsenal of sedatives and anaesthetics that allow the transformation of invisible hazards into progress, thereby tickling them from their slumbers. Competence is trumps. Tricks are won by involuntarily proving the contrary. 'Maximum safe levels', usable only in the plural, declare the opposite in and by their plurality. The more proofs that are presented of the official absence of hazardousness, the more powerfully they prove that they never were proofs, and never will be.

Whoever suggests that contamination in Germany after Chernobyl was 'not hazardous' thereby confirms that 'safe' nuclear power plants can so contaminate everything that creeps and flies over a range of several thousand kilometres, that only a medical diagnosis of their non-hazardousness can restore 'normality'. In order to declare that the experts have a firm grip on everything, these are sent out in the form of questionnaires; and, with an air of finality, they pronounce their self-cancelling, hydra-headed 'truths'. In the end, the official hazard-tamers themselves are locked in the cage of their own euphemisms; they are themselves the monkeys into which they aim to trivialize human fears.

To the turnover of hazards, become independent, there corresponds an unmasking, become independent, of the prevalent philosophies and technologies of safety; the latter are compelled to renege on their pledges, precisely as their claims are jeopardized. It may be that the nuclear power plants are becoming safer and safer. Yet those who run them and construct them, and the foundations on which they stand, are in each case becoming more and more insecure.

Taken together, both movements give rise to situations of clear danger, with ambivalent-explosive political potential. The mixture of political gases brewing up here can be ignited by the shower of sparks from the next accident.

Large-scale hazards: a 'revolution' without a subject

Hazard is not a revolution in the customary sense of the word, since the prevailing conditions themselves produce it; property and power relations remain constant, politics are unchanged. Thus it is an involuntary revolution; a revolution without revolution, which the prevailing conditions have instigated against themselves with the objectified power of the industrial momentum, and cannot now stop. In the sociological picture of hazard, painted here in broad brush-strokes, the conceptual relations of Marxian revolutionary theory are accordingly turned upside down. Instead of the coming to consciousness preceding action, the order has been reversed. Action is perpetually going on; it is the danger that changes the world. Consciousness limps a century behind the deed, as Günther Anders vividly puts it. The gap of a century between the deed and its representation also means that the deed does not happen because it has been legitimated, but because it is unrepresentable. Only the representations of our deeds are (perhaps) legitimated; this, too, only in the confusion of one century for another.

The actual revolutionary deed is thus not the deed, but the creation of awareness of the autonomous revolution of hazard that industrialism has turned into in its phase of technological self-creation. 'Don't act – think it over' is the motto. Overturning the relations of thought becomes the goal of a revolutionization of consciousness, which catches up with and recovers the revolution of self-sufficient action. It is not a change in the relations that is aimed at, but a change in the change: more precisely, a cutback in the self-sustaining hazard; a throttling down of action to an imaginable, controllable and responsible level. For that, however, there must be a change in the situation.

This change within change does not implement itself in the ascent from class *an sich* (in itself) to class *für sich* (for itself), from being to consciousness, but precisely the reverse. What happens anyway is raised into consciousness, and thereby exposed to the institutional pressure on it to change. The knowledge of its occurrence is decisive. The inversion of deed and consciousness also means that once hazards are socially acknowledged, the wheels of state bureaucracies are put into motion. In view of these upside-down relations between revolution and consciousness, Marx's old version of Feuerbach must surely be stood on its head once again: society has been changed to the point of unrecognizability; now the point is to reinterpret it.

In every theory of revolution, the search for the revolutionary subject begins here. The social sciences have not yet called off the search, fruitless though it has been. One is reminded of Bertolt Brecht, whose natural scientist Ziffel in *Flüchtlingsgespräche* ('Refugee conversations') says: 'That's how the proletarian comes by his mission of raising humanity to a higher level'; a short while later, the worker, Kalle, says: 'I thought so. The proletarian is supposed to be the carthorse again' (Brecht 1967, p. 1441).

If there is one characteristic that distinguishes the social dynamic of annihilation hazards from revolutions, it is the historical novelty that these play the role of 'carthorse' themselves. Certainly attentiveness, activity, social commitment are indispensable. However, the role of 'revolutionary subject', in the form of hazard, can become just as objectified as the relations whose abolition it carries out, wherever the social movements have sounded the alarm. The 'revolution' of hazard is carried out by the situation and its agents via its denial, in a kind of involuntary moonlighting – more thoroughly and impressively than anything the short arm of social movements could ever achieve. Self-annihilation hazards are the special case, never fully thought through, of the identity of the subject and object of the revolution. In various ways, the subversive counter-stroke is also perpetrated, with involuntary self-sacrifice, by those against whom it is directed. Nothing encourages hazard so much as its concealment. Those who conceal hazard secretly foster it. In the end, the hazard, together with the objectified violence permitted, encouraged, by its being interpreted away, can prosecute for the action that begins the change of change.

The hazard, too, is externalized, bundled, objectified subjectivity and history, a kind of 'solidified spirit', not unlike the ghost in the machine. While the machine bundles together various ends, hazard bundles together unperceived side-effects. It is a kind of compulsive, collective memory – of the fact that our decisions and mistakes are contained in what we find ourselves exposed to; and that even the highest degree of institutional independence is nothing but an independence that can be revoked, a borrowed mode of action which can, and must, be changed if it means self-jeopardization. Hazards are the embodiment of the mistakes of a whole epoch of industrialism, and they pursue their acknowledgement and correction with all the violence of the possibility of annihilation. They are a kind of collective return of the repressed, wherein the self-assurance of the industrial technocracy is confronted with the sources of its own errors, in the form of an objectified threat to its own existence. If one wanted to build a machine for abolishing the machine, one would have to base the construction on the hazard of self-annihilation. It is reification crying out for its annulment.

In this context, it is perhaps no exaggeration to say that the age of automation has created for itself, in the social mechanism of the hazard of self-annihilation, the sole possible 'counter-automatism' that is capable of abolishing in itself the automatism of the situation. To the detachment from the revolutionary subject there corresponds the liberation of the wage labourer from the means of production, with the admittedly substantial difference that it is the cultural counter-automatism of hazard that carries out the role of action. It is precisely the ambivalence of hazard, its integration into opportunities, wealth, technology, rationality, that grounds its objectivity, its independence of subjects. Hazard was made into the basis of social development under the guise of progress. This institutional internalization, this insider role, establishes it and frees it from subjective intervention. Even those who have driven it ahead, cannot now part from it overnight. Such a divorce would be outside the competence of any divorce court. The only counter-measure is denial. But that only fattens up the hazard even more. The protest can grow feeble, but the scandal of the hazard remains.

It may be true that hazard has taken on the role of an objectified third party, become independent, in the to-and-fro of public, political and ecology movement critiques, and that it is very adept at attracting attention in the age of the media. Yet this only means that the hazard performs a kind of objectified overture, providing many windows of opportunity for action, but does not thereby render action superfluous. On the contrary: the hazard constitutes an empowerment to act, and contains signposts for action, if one knows how to decipher them. It carries out the politics of detechnocratization, though one also has to follow up its tracks (chapter 7).

There is something particular about the tie that binds one to the revolutionary subject; its legitimation, in contradiction to the conflicts thus conjured up, has always been problematic. Think of the many violent measures that have been, and are, disposed of and implemented in the name of a 'dictatorship of the proletariat'. What is disputed is only whether the hazard is real, never its consequences (i.e. once its reality has been granted). It is open to us to doubt whether we are conjuring up self-annihilation, with the prevailing modes of nuclear and chemical industrialization. If this were recognized, however, then the radical character of self-annihilation would compel consensual action (whatever that may come to mean). Precisely because this is the case, the preventive prolongation of doubt affords the only chance for ensuring inactivity, for carrying on as before.

Thus as the hazard increases in magnitude and avoidability, it assumes the character of a self-legitimating evasion which, in the limit case, includes everybody it threatens. Thus, in the case of the danger of self-annihilation, the (potential) total mobilization contrasts with the selectiveness of revolutionary class consciousness. Conflicts with a logic of their own – mediated through epistemological filters and variables – arise here; conflicts about progress, which can no longer be contained, calling into question the hitherto prevailing conditions for their compensation.

The first main axis of conflict, intensified by the exacerbation and talking down of hazards of self-annihilation, is that between policy and administration on the one hand, and the public domain and its citizens on the other. Hazards dramatically undermine 'bureaucratic rationality', thereby opening gulfs between state authority and the democratic awareness of citizens.

'It's enough to turn you Catholic!' a committed feminist blurted out in a dispute about the consequences of human genetic engineering. This is precisely the point: previous political categorizations and fields of ideological force will dissolve in future conflicts about progress, and even social-industrial arch-enemies will be forced into 'rainbow coalitions' (cf. chapter 6).

To summarize the foregoing sketch, one can say that not only humans but also the bureaucracies are caught in the trap the world has become. This identification with one's oppressor may afford scant consolation. The insight may render the lack of an overall view even more confusing. Yet it is worthy of analysis. We need a quite banal social theory that illuminates not only individual frustrations, but also the 'paradoxes of the apocalypse' (Milan Kundera) in which the systems and institutions of the civilization of large-scale hazards have become entangled.

Contradictions of the security state: hazard as a refutation of the technically oriented conception of safety

In this section we shall address the question of how nineteenth-century industrial society learned from handling its consequences and dangers, and why risk society at the end of the twentieth century is threatening to forget the lesson. We shall do this through a reading of *L'État providence* (1986) by François Ewald, a follower of Michel Foucault.

In many schools of thought concerned with industrialism, the social process of rationalization is conceived essentially as an act of 'imperialist occupation'. That is to say, a view is developed of an industrial capitalism that liberates vast reserves of productivity, but, in developing this wealth, pillages the traditions it finds to hand (cf.

Habermas 1986–9; Beck 1983; Lutz 1984; J. Berger 1986). The viewpoint according to which the industrial epoch has created substantial institutions, that are themselves being undermined and questioned in a new phase of modernization, is only gradually gaining ground. This question of the creativity of industrial capitalism, and also of its self-destructiveness as measured against the institutions that were invented and implemented under it, is necessary for an adequate comprehension of the social interplay of safety and hazard in the nuclear age – and it is among the issues addressed by François Ewald in his book.

Ewald presents a study partly oriented towards jurisprudence and the philosophy of the state, and partly comprising a wealth of historical materials. This allows one to comprehend the production of security as a sociological phenomenon, based upon institutional inventions and arrangements with which industrial society deals with the dangers of its own creation.

From this perspective, risk calculations and private and state insurance policies are social answers to the challenge of the insecurities created by modernity in every area of life. Safety issues are thereby detached from the Procrustean bed of the individualist calculus of utilities (as in Schäfer 1972 and Kaufmann 1973), and from that of the technological minimization of accident probabilities; then they are subjected to the social organization of prior provision, responsibility and compensation.[4]

Conversely, these institutions for promoting the safety of industrialism from itself, laboriously negotiated and realized in social struggles, are endangered by the new generation of 'man-made', socially produced disasters. It is not (only) the outbreak of disasters, but even the mere threat of them that tears apart the institutional fabric of calculated precaution. In the epochal difference between early industrial risks and large-scale ecological, nuclear, chemical and genetic hazards at the end of the twentieth century, the technological security state enters into contradiction with itself. The lack of precaution, its impossibility, becomes a destabilizing force within the institutions, undermining and cancelling the 'insurance contract', which was invented by modern society in order to absorb the insecurities of its own making. Society ends up under the perpetual compulsion to renegotiate the conditions of its 'social rationality' at every level.

How, precisely, is the social construction constituted, via which large-scale technological hazards and natural despoliations are transformed into social dangers to the system? Ewald argues that the industrial epoch, with its dialectic between risk and assurance, has invented, thought through, and in social conflicts implemented and generalized a specific mode of societal management of the consequences of societal development. Industrialism is an adventure, a revolution of normality, continually producing incalculable consequences and damage. In the early phase these are blamed on individuals; only with the advent of the calculus of risk, in which particular circumstances are left out of consideration and accidents are calculated according to the probabilities of their incidence, does one attain the level of the social definition of consequences which corresponds to their aetiology. Risks turn events, initially foisted on individuals as their own errors, into social events, systemic events, which also require social, institutional, political regulation.

The epoch-making achievement of the concept of risk and assurance consists in a double socialization of responsibility and hazard. The threat is divested of its individual history according to the yardsticks of 'probabilistic reason', and the question of compensation is detached from questions of individual error and submitted to a general system of compensation. If there is no obvious culpability or gross negligence, then the consequences alone lead to their being compensated for, at least in principle. The advantages of this institution of the compensatory anticipation of consequences, moreover, clearly lie in their (stochastic-statistical) ability to calculate the incalculable (with the help, say, of accident statistics); the abstractness and universalizability of their formulaic solutions; and their general economic compensation, predicated on the principle of money in exchange for destruction.

To that extent, the idea of insurance, which appeared on the threshold of the modern era as a means of counteracting the uncertainty of marine trade routes, is a 'daughter of capital'. It spreads as traditional forms of solidarity dissolve; it follows in the tracks of the insecurity brought into every nook and cranny of existence by modernity; and it creates, through private and public insurance contracts, a thoroughgoing social relation, a kind of 'social pact' located somewhere between liberalism and socialism. The consequences are extricated from the categories of good and evil, culpability and error, and fed as statistically calculable 'destinies' into social rules of restitution. In this manner it becomes possible to attribute the unattributable, to calculate the incalculable and to generate present security in the face of an uncertain future. The threat represented by modernity is defused in the present by an institutional anticipation of its consequences.

To that extent, the calculus of risk and the idea of assurance represent more than just one institution among the others of industrial society. They are the institutional arrangement, the invention, with which industrial modernity anticipates and compensates the insecurities of its own creation. In this way the 'assurance state' (*état providence*) arises as a pendant to 'risk society'; a social architecture whose social integration, idea of justice and social contract become sociotechnically shapeable and perfectible, in line with the social rationality of the administration of hazards and consequences, i.e. according to the model of risk and assurance.

Large-scale nuclear, ecological, genetic and chemical hazards break in at least three ways with the established logic of risk. In the first place, they involve global, frequently irreparable damage: the concept of financial compensation fails to apply. Second, prior provision for the worst conceivable accident is out of the question in the case of annihilation hazards: the idea of security through anticipatory control of the consequences fails to apply; it is replaced by contradictory, infinite technological security and the dogma of 'residual risk'. Third, the 'accident' loses its (spatio-temporal) limitations, and thus its meaning; it becomes an 'event' that is forever beginning, an 'open-ended festival' of creeping, galloping and overlapping despoliation: that means, however, that norms, the foundations of measurement and thus of hazard calculations, cease to apply; incomparables are compared; calculation turns into obfuscation.

This legalization of incalculable, unrestitutable hazards of annihilation, however, exposes the social contract of the security state to erosion from within – (relatively) independently, moreover, of physical-technological devastation. The 'limited resonance' of the subsystems (Luhmann) is only one side of the contradiction in which the institutions end up.

In the developed security state, these naturally mediated dangers to the system assume the form of a 'crisis of responsibility' (Ewald), compelling one to rethink the problem of attribution and regulation in society.

Where progress and doom are interwoven, the goals of social development are spelled out contradictorily from top to bottom. Certainly, this is not the first conflict that modern societies will have to withstand – but it is one of the most fundamental. Class conflicts and revolutions change power relations, replacing one elite by another. Yet they cleave to the goals of technological-economic progress, and take place within a dispute between shared notions of human and civil rights. The Janus-headed 'progress of self-annihilation', on the contrary, gives rise to conflicts that put into doubt the social basis of rationality – science, law, democracy. Society is thereby coerced into perpetually negotiating its basic principles, without basic principles. It must undertake to know itself without epistemological foundations; it falls prey to an institutional destabilization, wherein all decisions – from the 'parking-space policies' of district councils, through the details of manufacture of industrial goods, to the basic questions of energy provision, law and technological development – can suddenly be drawn into the slipstream of political conflicts about principles.

Where the façades remain intact, power positions akin to subsidiary governments arise in the milieu of definition- and publicity-dependent hazards: in the meteorological offices, research laboratories, nuclear power plants, chemical factories, editorial offices, lawcourts, etc. To put it differently, as the contradictions of the security state are fanned into flame, the apparently 'self-referential' systems become susceptible to actions and subject-dependent. The courageous Davids get their chance against this Goliath. The colossal interdependence of definitions of hazard – the collapse of markets, property rights, union power and political responsibility – leads to the emergence of key positions and key media for the definition of risk, cutting across the political and occupational hierarchy (cf. chapter 6).

The relations of definition, the system of organized non-liability, are increasingly moving to the centre of the critique (see chapter 7). Anyone who succeeded, for example, in establishing new rules for the causal attribution of hazards to the international traffic in pollutants (that is, in enacting a 'reform of causality') would have carried out something like constitutional change, and by that alone created a different state. For this would induce profound changes in the economic-technological and legal fabric. It is not a refutation of this assessment that appropriate opportunities for intervention would first have to be struggled for, cleared in debates and secured in law – but proof of it.

Implementation as Abolition of Technocracy: *The Logic of Relativistic Science*

Let us suppose that the case of Galileo is reopened: natural science against the Church, heard in 1633 by the Holy Office in Rome. The Grand Inquisitor orders a retrial on the grounds of Chernobyl, human genetics and the rest. What would the verdict be today?

Only a decade or so ago, every schoolboy's heart beat faster for the unyielding Galileo: 'And yet it does move!' That was the dream of the liberating power of science. Brecht's *Galileo*, written in exile in 1939, utters the bitter truth: 'Take care of yourself; when you go through Germany, put the truth under your coat' (1967, p. 27). The world may be in disarray, but the world view is correct. Truth equals science, and science is the hope of a world martyred by fascist madness and world wars. Even today, some scientists have to carry the truth under their coats – in order to keep the public from getting too scared.

Are the truth and foresight of the Grand Inquisitor beginning to dawn on us, living as we do after Nagasaki, Bhopal and Chernobyl, and before the ultimate victory of human genetics? Have we changed sides, behind our own backs, so to speak, in the centuries-old dispute between science and dogma? Or is it science that has changed sides, so that the fading cry of 'Help, science!' is ambiguous?

Are we now also being liberated from secular faith in science and technology, as people in the age of the Reformation were 'released' from the worldly arms of the Church? Are we experiencing the first waves of a scientific Protestantism in the public stands of critical nuclear scientists, geneticists, doctors, engineers? Will there be a great dispute over the vested interests of risk-intensive research ventures, like the dispute over the 'sins of the Church' which convulsed the late Middle Ages? Are the old spectres of irrationalism, long since pronounced dead, rising from their graves, and, now immune to a scientific critique that has deconsecrated itself, beginning to haunt us anew?

It is a mistake to throw these questions into the waste-paper baskets provided in our day for anti-technological rantings. Certainly, the music of world decline has accompanied the development of technology before now: a great deal sounds as it did 200 years ago, and frequently since then: sentimentalism about nature, alternative lifestyles, Luddism, etc. Yet today it is not enemies of technology, apostles of a new stone-age culture, who warn against developments, but specialists who know their way around and have courage. The situation is changed by the fact that the critique of technology and science is also domestically produced, and thus able to appear with scientific rigour and authority. The public self-refutation of scientific ideals of certainty does not indicate a failure of science, but a new phase in its development.

The reign of the refuted: technocracy and the 'state of the art' principle

We live in a second-order reality, constructed out of science and technology. Reality has become an unstable product of the spirit of the laboratory. This imbues science and technology more deeply than ever with the promise of salvation, as is surely nowhere so apparent as where both are damned and praised at once as poison and antidote. Against the errors of scientific rationality, what is prescribed and demanded everywhere is scientific rationality – only better, deeper, more thorough, more alternative. Ecology is starting to blossom into a superscience, for which earlier errors are becoming rungs on a ladder to more and higher mathematics and formalization. It is thought necessary only to dive deeper into the sea of uncertainty, in which science itself has plunged everything, in order to find somewhere the sources of security on earth. Anyone who blames everything on science and technology, seeing a world wedded to its future apocalypse, has a couple of figures available as 'proofs' – as a source of faith, from which his or her pessimism can draw its nourishment. In the technologically constructed world we are condemned to scientificity, even where scientificity stands condemned.

This impossibility of a scientific vacuum in risk society is also evident from the coincidence, in situations of extreme hazard, of radical questioning and scientific dependency. An extreme example may show how far everyday life falls under the spell of the scientific diagnosis: the children in settlements near toxic waste dumps learn the chemical formulae before going to kindergarten.

> Oliver kneels by a puddle, and hacks the ground with his child's shovel. Soil flies through the air; he runs to the back of the house, fetches rubber gloves and spades, takes dug-up soil to his 'transit camp' behind the garage, and continues digging.
>
>   'Oliver, what are you doing there?'
>
>   'Getting poison.'
>
>   'Do you really believe there's poison there?'
>
>   'There, poison – black!' The soil really is a greenish-black colour.
>
>   'What kind of poison are you digging up?'
>
>   'Lead, cadmium, arsenic, cyclopentadia.' Five-year-old Oliver, already an old hand with poisons, has a better command of chemical jargon than some corporate chemists.
>
>   The toddler explains in a piping voice: 'Because it's poisoned here, totally poisoned, the whole settlement is totally poisoned . . . they have a sign up, too, and I can read it now: En-trance pro-hib-ited.' The truth is, he cannot yet read. He knows the signs, and knows that he is not allowed to play there. His mother has dinned it into him.
>
>   We are in Dortmund Dortsfeld-Süd, a 'poisoned estate' . . . we have shot our first footage of a child 'playing' at detoxification: searching for poisons, testing them and rendering them harmless – with a little transit dump before they are finally dumped. Ever since Oliver could hold a spade and a shovel, he has almost compulsively played at 'detoxification'. Chemicals specific to the coking process, from dicyclopentadia to the carcinogenic benzol, determine the everyday life of this settlement, including the children's. (Kerner 1987, p. 38)

Hazard situations are a dilemma in which individual scientific assertions turn involuntarily into their own refutation. It is as an unequal trinity of originators, diagnosticians and agents of salvation (under suspicion of the contrary) that scientists and technologists are involved in them. It goes without saying that chemists, nuclear physicists, genetic engineers (let us call them A) took part in the creation of hazards. It is equally clear that these sciences and scientists must put their concepts and methods (B) at the disposal of society, in order to enable an assessment of what threatens everybody, but is perceived by none. B refutes A. The successful demonstration of hazard refutes the original safety declarations. For all the investment in mathematical ceremony, the more recent claims thereby lose some of their immediate plausibility.

Nor is the only contradiction in this process the short-circuit of counter-arguments, wherein science in hazard situations closes upon itself, behind the backs of upright scientific specialists. On the one hand, technologists claim that their declarations about risks are strictly scientific, including the maximum permissible levels on which they depend. The observer rubs his eyes in disbelief at the miraculous power of the engineering sciences; these have now managed to overcome the gravity of the insight that no force in the universe, no matter how sparkling its formulae, can yield normative propositions concerning risk acceptance. Yet no sooner has one accustomed oneself to this feat of magic than one has to acknowledge that the technologists have done more, hanging the norms they claim to prescribe from the gallows of fluctuating maximum permissible levels. Now the technologists' claims to accuracy dangle, one beside another, from the rope of the maximum permissible levels they have tied around their own necks.

In a certain sense, we have here a case of the revenge of the side-effects upon their creators. In its involuntary (synchronic and diachronic) contrariety, science amounts to the contrary of its claims.

> Three times in the last thirty years we have experienced the fall and reinstatement of aspirin; child nutritionists have declared mother's milk to be worse, better, worse, etc. than chemical products; valerian was thrown out of most pharmacopoeias, and then brought back. A good memory is a strong argument against the claim of a scientific result to be 'true' now and forever. (Löw 1984, p. 401)

Politicians, too, can sing a song of the transience of scientific proofs. Where the legalization of animal experiments is concerned, experts hygienically excise every doubt, painting a pitch-black picture of the dangers to the health of the population that a restriction on animal experiments conjures up. Next door, a member of parliament is briefed by the advisory committee for legislation on animal experiments. Always receptive to good arguments, he or she is minutely and irrefutably apprised of the fact that experiments on rats, mice, pigs, etc. are of no relevance to human beings, and that humanity can therefore only be protected from the worst by giving carte blanche to experiments on human embryos.

There is nothing new in the relationship between expert opinion and counter-opinion; nor does it subsist only mediately through public opinion, in exchange with social movements. Ever since it began, science has been an undertaking to refute itself: in the competition over progress, research financing, professional reputations and markets, science refutes science – with that thoroughness and demonstrability of which only science is capable. This effect becomes more systematic and apparent with the increasing differentiation of science; with the speed and complexity of its development; and in so far as the sciences extend their questions to their own consequences, errors, and foundations.[1]

Yet whoever sees this dare not say it aloud, for the contrary soon catches up with him: the abstract (un)reality of the hazards has played the monopoly on their interpretation into the hands of their creators. In the first shock of disaster, everybody talks about Becquerels, rems, glycol, as if they knew what that meant; and so they must, in order to find their way around the most everyday matters. After the dead certainty of the next non-accident, there will no doubt be professorships in infant nuclear anxieties or radioactive cabbage: refutation equals expansion. Hazard compels the return of faith.

This can also be seen from the fact that police measures against hazards are increasingly being replaced by the risk assessments of engineering science: that is shown not only by the professional expertise upon which political, economic and administrative decisions are based, but also by the manner in which critics of modern technology attempt to substantiate their claims in the courts (cf. Wolf 1987).

Nearly every misgiving about the risks of a technology is formulated on the basis of dominant scientific-technological thinking. The therapies and alternatives on offer also nearly always derive from the arsenal of technology. Objections to conceptions of nuclear power plant safety are predominantly based on the critical insights of reactor technology itself. Desulphurization techniques are demanded to counteract SO_2 emissions, and catalytic converters are supposed to prevent damage from exhaust fumes. An appeal to the legislator usually demands no more of the concepts invoked, from waste recycling to precautions against nuclear plant explosions, than that they should send out environmental policy signals for a supposedly improved technological solution. And many of the charges preferred in court, according to the plaintiffs' own understanding of them, demand only that the findings of natural science be legally enforced. Maximum safe levels are exposed to doubt if there are indications that the underlying evidence for them is outdated, or that measurement procedures are technically obsolete. Emissions from old plants come under criticism if those affected suspect them of being above unavoidable minimal levels, according to the 'state of the art'.

This is the contradiction that must be exposed: on the one hand, the contradictions between the individual claims of the technological sciences mean that the latter inadvertently carry out their own self-refutation. On the other hand, for the past century they have uninterruptedly enjoyed the privilege of deciding, in accordance with their own internal yardsticks, the paramount social and political question: how safe is safe enough? One of the central pillars of technocracy has become unstable here.

In the process, the technological sciences have anchored their society-creating role in a simple social construct. It is granted to them to make legally and politically binding decisions, by their own yardsticks, as to what the 'state of the technology' enjoins. But since this general formula is the yardstick for actionable breaches of safety, private organizations and bodies, indeed, just a handful of experts, decide *de facto* on the hazards to be foisted on everybody (Rossnagel 1983; Wolf 1986a; Roters 1987).

Power over the definition of the key cultural questions of human genetics and of 'medical progress', as the insider Jacques Testart attests, is also concentrated in the hands of a few. 'Medicine has privileges unlike those of any other profession . . . Whether in the field of cancer or Aids research, in every case there are ten to fifteen people who have the last word. And they really do as they like; they even lay down the laws, since the ministers take their advice' (Testart 1988, p. 66).

Technological hazards that have shed the armour of their official denial convulse governments, lead to the collapse of markets, neutralize borders, cause the world to shrink to a locale whence there is no escape. If, however, one pushes aside the façades of liability and forges ahead to the small print of safety standards, it is consistently the case that in all the central issues – from reactor safety, through the prevention of air, water and noise pollution, to the medical shaping of life – it is neither parliaments nor governments nor lawcourts that wield the power, but technologists and physicians.

If one asks, say, what levels of artificially produced radiation the population is to accept, i.e. where to situate the thresholds of tolerance that separate normality from danger, the *Atomgesetz* (Nuclear Law) provides only a general answer: the requisite prior provision is to correspond to the 'state of science and technology' (para. 7 II, no. 3). This formula is then filled out in the 'guidelines' of the commission for reactor safety – an 'advisory body' of the Ministry of the Environment, in which representatives of the engineering associations have the last word (Wolf 1987).

The pattern of policy-making is always the same, whether in the fields of air, water or noise pollution. Laws prescribe the general policy conception, but in order to know the permanent ration of normal poisoning thrust on every citizen, one has to pick up the EC Large Combustion Plant Directive, the German air quality guidelines or the British 'Notes on Best Practical Means', and read up the crucial details.

Even the classical tools of policy control, legislation and administrative regulations, are hollow at their core, juggling with the 'state of the art'; in this way they subvert their own responsibility, simultaneously replacing it on the throne of hazard civilization by 'scientific-technological expertise'.

This delegatory formula of the 'state of science and technology' has a lot going for it. Its apparent self-evidence is exceeded only by its political significance. It represents as it were the revolving door between democracy and technocracy. Everyone thinks it self-evident that technologists must have the last word on issues at the 'core' of technology. This is valid if only because the breakneck pace of technological development allows for no enduring codifications, and cannot be comprehended by outsiders. On the other hand, the engineering associations, the medical geniuses, etc. are thus given carte blanche to make fundamental political decisions. The democratic institutions sign their declaration of surrender, and in the splendour of their formal responsibility they delegate power over matters of safety to the technocratic 'alternative government' of corporately organized groups.

To this extent, the 'state of science and technology' formula amounts to an implicit enabling law. Government, parliament and jurisprudence disempower themselves and, in so far as the repercussions of technology change society and determine the political agenda more than any proposed legislation, the organs of the state become branch offices, executive bodies, mere apologists for the advisers' and engineers' associations which actually govern by donning the mantle of the 'technically necessary' and wielding the sceptre of 'expertise'.

Thus we are dealing here with a core structure of technocracy, which establishes the relations of definition (i.e. power relations) of safety issues above and beyond the limits of the subsystems. Just as property relations transfer to the owners of capital the disposition over the character of labour and distribution of profits, so the 'state of science and technology' clause enables the institutes of technological standards to make binding decisions on highly political safety matters. Precisely therein lies the fundamental technocratic contradiction in the atomic, chemical and genetic era: the spatial and temporal extent and the threat of its consequences are becoming unimaginable, the very borders of national states and of military blocs are being annulled by what threatens; and there, in a blossoming democracy and with the blessing of its fully functional institutions, a small group of corporate engineers and physicians actually decides the conditions under which people live, in the multiplicity of states from the Urals to the Atlantic.

It is instructive to inquire into the historical genesis of the sciences' power over definitions. For the technologists did not seize power in a *coup d'état*. Nor do we live in a technological dictatorship. Rather, the increasing range and political quality of the safety issues thrown up by technologies has led the power relations to shift towards a technocracy. At the start of the twentieth century, in the early phase of late industrialism, it seemed plausible and logically sound to leave questions of safety to the technologists. In the transition to the atomic age, this has blossomed into the political privilege of defining, bindingly for one and all, the conditions for survival in scientific-technological civilization. The law and its legislative paradigms have not followed the historically changed facts that they regulate.[2]

The heightened version of scientific knowledge: truth, doubt, self-limitation

In *Dialectic of Enlightenment*, Horkheimer and Adorno write (and this is a substantial reason for their profound scepticism): 'Science itself has no consciousness of itself, it is a tool . . . to conceive of science as understanding itself runs counter to the very concept of science' (1979, p. 85). On that point they are entirely in accord with their arch-opponents, the optimists of progress (e.g. Francis Bacon or Auguste Comte): above and beyond all the theses and antitheses, it is only the rationality of the claims of the scientific attitude that is under dispute, never their efficacy. Critical theory sees science and technology turn into the contrary of their claims, because the law of their development appears to be unstoppable, and because everything that might have constituted a counter-movement – ethics, local protest, public opinion, the oppressed class, politics – either finds itself compelled by science to dissolution, or must quietly desert its ideals in order to put in an appearance.

Thus the distribution of roles is established in a concerto for antiphonic voices. Cassandra is played by an employee in that ancilliary service of science, ethics. With her clothes torn to shreds by science, herself lacking foundations, she must cry out her criticisms to the audience. Even Bacon, whose naive optimism of reason elevated happiness through science into a programme, conjured up the guardian role of ethics, the scientists' oath not to reveal knowledge, in order to conceal what should be kept secret.

In the model of the objectified sciences, ethics plays the part of a bicycle brake on an intercontinental aeroplane. Certainly, progress without any moral horizon would not breach any norms, and would thus stride forward impervious to all critique. But it is due to the ineffectiveness of a merely ethical critique that it can be given a high polish. As is well known, the eternal round of points of view amounts to a graveside lament for ethical obligation. Thus one need only bring one more question into play, for the public clamour in the antechambers of science to busy itself with – itself. We, on the other hand, intend to inspect the land along its fault line, in its holy of holies, the logic of discovery.

Besides the many deeds that it celebrates, science has carried out one that it keeps secret even from itself: the true–false positivism of clear-cut factualist science, at once this century's article of faith and its terrifying spectre, is at an end. It continues to exist only as a narrow, specialist consciousness, although as such it is admittedly very effective. It is something of a sorcerer's apprentice, in having itself succumbed to the doubt it let loose.

If one poses an arbitrarily chosen question to a motley panel of scientific experts – say, 'Is formaldehyde poisonous?' – one will be given fifteen different answers from five scientists, all of them garnished with various qualifications – if, that is, the persons questioned are good at what they do; otherwise, two or three superficially unambiguous replies. This is neither by chance nor by accident. It represents the state of science at the end of the twentieth century. The unambiguity of scientific statements has eluded insight into their dependence on decisions, methodology, context. A different computer, a different institute, a different employer: a different 'reality'. It would be a miracle, and not science, were it otherwise.

Technology and natural science have become one economic enterprise on a large industrial scale, without truth or enlightenment, comparable to the secular power of the medieval Church without God. Just as the Church of the Inquisition failed to prove the existence of God, so the reigning science still owes us proof of its truth. Moreover, it has involuntarily and indirectly furnished the proof of its contrary, across the length and breadth of its victory parade.

Those who set their doubts upon their own foundations, for a change can discover for themselves, if they do their job well, that these foundations do not exist. Thus the experimental grounding which is at the very heart of the natural sciences has quite simply been demolished this century in long, thorough debates. The only cure is to overlook the fact. Back to business, in the Nirvana of details!

Anyone who swears by specialization ought not to be surprised if all publicly disputed questions end up with incommensurable results; frequently, these do not even contradict one another, simply because they cannot be brought into relation. These are all successes due to the implementation of scientific claims. In the end, doubt even vanquishes the episteme to which it owes everything. That not only gives cause for tears, it also gives the scientific monopoly on knowledge a subversive democratic undertow. Now others can join the discussion. A few basic methodological objections, to be on the safe side, and undesirable results will collapse. Reset the points, and the train of argument will go off in the opposite direction.

There is an analogy between the development of (natural) science and the decline of Catholicism before the Reformation. Faith – respectively in God and in science – is undermined by the ritualism of institutionalized practices in open contradiction with what they declare. In their day, too, the princes of the Church reacted like today's princes of science – with cynicism. Dostoevsky's parable of the Grand Inquisitor, a cunning, cynical power-monger, could be translated without much distortion into the scientific milieu of today. The silenced quest for the truth everywhere comes up against routine, career and, as a last justification, either naivety or a cynicism that denies itself to be such: that would be quite a position!

God and the Church have been replaced by the forces of production, and by those who develop and administer them, science and the economy. The actions of the latter bear all the hallmarks of religious belief: trust in the unknown, the unseen, the ineffable.

Progress is the inversion of rational action in a 'process of rationalization'. It constitutes a licence for a permanent unplanned change of society, without the latter's sanction, towards the unknown. It is supposed, for instance, that the genetic tinkering in prospect can ultimately be turned to good. But even to ask how is somewhat heretical. Assent is presupposed, without knowledge of ends. Anything else is misguided.

Art too has outstripped science in its claim to knowledge. As early as the beginning of the twentieth century, Cubism, for example, dissolved the fictitious object into the multiplicity and polysemy of perspectives on it – while the apparent unambiguity of an empiricism still unconscious of its constitutive role continued to prevail in the natural sciences. It is only today that the sociology of the family acknowledges (albeit reluctantly) the decline of the bourgeois ideals of marriage and the family, as represented on the stage by Ibsen and Strindberg; now it can no longer be hidden among the statistics. The fragmentation of the omniscient author into the consciousnesses of his or her characters, which has given wings or leaden shoes to the figure of the narrator since the days of Joyce, Musil and Thomas Mann, while compelling the narrator to turn the production of the story itself into one of its themes – that is still powerfully repressed in science. Data are certainly manufactured products; theories, no less than novels, are tenuous constructions. But the author of scientific penny dreadfuls shelters more than ever behind the now transparent fiction of scientific neutrality and routine.

In a bravura display of the scientific method, the sciences extend their own logic to convict it of illogicality. It is not the scientists who practise self-criticism, but science itself. It is an involuntary form of self-mortification, availing itself of the research procedures of individual scientific fields. The critique of science subsists in context, enacting itself behind the backs and above the heads of individual researchers.

What is it? Revolution? Change? Crisis? Loss of legitimation? All and none of these. It is the hitherto unperceived relationship between the implementation and the abolition of scientistic ideals of knowledge. This relationship is not obvious; nor does it declare itself. It must be uncovered and made public. But it is present everywhere, as a state of objective morbidity, in the conflict between individually 'valid' scientific results.

If it had been assumed, with naive faith in science, that epistemic progress was a cumulative approach to truth and security, that increased scientific research leads to the greater unassailability of its pronouncements, then this inadequate image of scientific work is turned upside down as the scientific society is established. The heightened version of scientificity is: truth, doubt, self-limitation. People who inquire further, learn more, and also more about the fragility and limitations of their own fundamental principles and statements; and therefore they learn less.

'Does that constitute an abandonment of the scientific ideal of truth?' asks Hartmut Esser (1987). The answer is no. 'Scientific rationality' has half dismounted, half been pulled down from its high horse, and it remains for us to re-determine and reclaim it. Karl

Popper was surely the last to find the logic of scientific discovery in its actual practice. Perhaps this is the borderline between the relativistic science of the future and its authoritarian predecessor: science has no logic. The logic of discovery is not a possession, a grail that can be safeguarded. Its logic is a project, a utopia that has to be continually regained in the struggle with historical challenges and refutations, but it can also be gambled away. The ability to learn about big issues, the logic of discovery and the establishment of fundamental principles; the ability to learn, as an institutional structure, also about the handling of hazards now only perceivable by science – that would be one of the essential magic formulae.

This also applies to the content of the social sciences, whose subject is hardly impervious to historical change. Where everything is submitted to a decision, the number of variables increases; and with increasing knowledge of the variability of the conditions, fewer and fewer facts can be stated with certainty (and for good reasons). Nor do multivariable models, huge databases or alliances between sociological research institutes provide a cure for it: to know more is to know less. To judge from post-war developments, that proposition was surely never so true as it is today – and will be tomorrow. Population growth depends (among other things) on the level of employment among women, and that in turn depends on the development of the labour market, but also on the technological policies presented by the government. However, one can also rotate this cross-section of possible 'second-order variables' around the axis of other decision-dependent parameters; the problem remains the same. As the potential for decision-making grows in every field (across the spectrum of private life, the economy and politics), the gap widens between the supposed conceptual-schematic order and the increasingly unforeseeable multiplicity of social developments. The uncertainty of the social sciences (assuming for once that their methods are sound) is thus substantially a function of the instability of social conditions. Where the latter slide out of control, a sociology that looks essentially to method for its real content threatens to become unreal. Every individual assertion, no matter how scientifically impeccable, threatens to drown in the questions to which it gives rise.

The world as a laboratory, or the end of experimental technology

The real scandal surely consists in the reactions of the natural and technological sciences to the hazards of their own manufacture. True as it was before Chernobyl, it is only since 1986 that a wider public has been apprised of the following: there are worlds of difference in the slight distinction between safety and probable safety. The technological sciences only ever have probable safety at their disposition. Hence their statements remain true even if another two or three atomic energy plants blow up tomorrow.

Wolf Häfele wrote in 1974 (and at the time it sounded innovatory):

> It is precisely the interplay between theory and experiment, or trial and error, that is no longer available to reactor technology ... reactor engineers take this dilemma into account by breaking down the problem of safety technology into sub-problems. For example, the solidity of the compression system is analysed as a component problem, as are the functioning of control and pump systems ... However, even splitting up the problem can make possible only an asymptotic approach to complete safety ... The remaining 'residual risk' opens the door to the realm of the 'hypothetical' ... The exchange between theory and experiment, which traditionally leads to the truth, is no longer possible ... I believe that this final indeterminacy in our undertaking partly explains the especial sensitivity of public opinion to the safety of nuclear reactions. (pp. 313ff)

What is touched upon here is nothing less than the contradiction between experimental logic and nuclear hazards. Just as sociologists cannot force society into a test-tube, so technologists can only test nuclear reactors if they turn the world into a laboratory. Theories of reactor safety cannot be checked before the construction of nuclear power stations, but only afterwards. The expedient of testing subsystems separately makes their interaction less predeterminable, thereby producing sources of error which, for their part, cannot be experimentally controlled.

Compared to the logic of discovery as originally negotiated, this represents a straightforward inversion. The sequence is no longer first laboratory, then application; instead, the examination follows the realization, manufacture precedes research. There is a Marxist core lurking undiscovered in the modern technological sciences. They have surreptitiously turned praxis into the criterion of its truth.

The dilemma into which large-scale hazards have plunged scientific logic holds for all experiments (i.e. nuclear, chemical and genetic): science hovers blindly over the brink of dangers. Test-tube babies must be produced, genetically manipulated beings engineered, reactors built, before and in order that their characteristics and safety may be studied. That is, the question of safety must be answered in the affirmative, before it can become an issue.

While a favourable verdict on safety is the gateway to every practical-empirical examination, research continues to depend on an ultimately favourable answer to the safety questions it throws up. For doubts about safety would mean the end not only of praxis, but also of research.

Thus research into the safety of highly dangerous systems is only possible if it is circular, that is, provided that its safety is assumed in advance to be assured. This circle is founded upon the mutual embrace of research and application. A no to safety in its practical realization is a no to research.

In the nuclear age (and this applies also to the chemical and biological revolutions) the danger of self-annihilation thus triggers a rapid drop in the exchange value of scientific plausibility, a kind of 'crisis of world science', comparable to a world economic crisis but even more insuperable. The end of experimental science coincides with the compulsion to set the hazards free in order to research them, and with the circular argumentation of a partisan science. Translated into legal terms, this means that the defendant is his or her own judge on matters relating to technological safety.[3]

The image of the sorcerer's apprentice acquires a concrete meaning: 'applied research', having been bridled from behind, ends up under the tyranny of the hazards it has conjured up. The latter dictate to their technological masters the conditions for research into them. They stand the logic of discovery on its head, force researchers to renege on fundamental principles, and leave 'scientific rationality', once so proud, to sign its declaration of surrender under the appalled eyes of world opinion, step by step, accident by accident. Underlying this is a chain reaction (over a period of several decades) whose individual components can be clearly discerned.

By anticipating the application prior to its discovery, science itself has torn down the dividing walls between the laboratory and society. That has led in the first instance to a shift in the conditions for freedom of research. The latter implies freedom of implementation or, to put it more strongly, carte blanche for research applications, without knowing their consequences, in hazardous fields that threaten humanity. Today anyone who demands or admits only freedom of research abolishes research. In the fields of nuclear reactors, genetic engineering and information technology, research is carried out under the Damocles' sword of colossal investments. Natural-scientific dogma thrives in the shadow of large-scale risks. Errors into which billions have been pumped can no longer be admitted.

The illogic of discovery

The crisis of the natural and technological sciences tends to be discussed today as a question of self-limitation, of ethics. But we must surely go one step further. The seemingly self-evident translation of natural-scientific procedure to large-scale hazards has covertly invalidated the foundations of natural-scientific experimental logic. What is at issue here is nothing less than a classic falsification (in the strict Popperian sense) of the logic of discovery itself.

1   Hypotheses about the safety of large-scale hazards cannot be tested experimentally. The greatest possible hazard invalidates the principles on which scientific safety procedures have been founded hitherto.

2   The world is turning into a laboratory. The experiments being carried out on humankind elude intervention from natural and technological science. Experiments and their outcomes have become undelimitable (as have accidents) – they have no spatial or temporal boundaries, being both international and interdisciplinary, their consequences and errors unattributable.

3   Natural science has thereby forfeited its exclusive right to judge what an experiment signifies. Research has, as it were, been implicitly socialized. Public opinion and governments, being components of the experiment, demand to have their say. Real-life experiments, such as the evaluation of nuclear, chemical and genetic hazards, have become fundamentally ambiguous and open to interpretation: the experience of the public breaks away from controlled scientific experience, and the two engage in a competitive struggle over interpretation of the results.

4   Just as they abolished the laboratory experiment, so large-scale hazards have abolished the accident – at least as a spatio-temporally delimitable event. Accordingly, the yardsticks by which the hazard's hazardousness is measured – the number of dead, wounded, etc. – derive from a different century, and the principles on which accident statistics are calculated are so much waste paper. The number of Chernobyl dead will never be counted. Most of them have not even died yet. Generations and countries knowing virtually nothing of the event will have to pay with their lives and their health for the long-term devastation, beyond human comprehension, wrought by this 'accident'.

5   The repercussions (perhaps refuting all the initial hypotheses) are dealt with interdisciplinarily, thereby opening the door to vari-

ous methods of calculation, miscalculation and recalculation. Specialist decisions, statistical procedures and definitions are significant factors in deciding whether and how far the unattributed repercussions are freed from compartmentalization (such as cancer, leukaemia, etc.). An investigation carried out by a doctor in Britain into the link between child cancer and radiation density is of comparable significance for the global nuclear experiment to that formerly accorded to laboratory tests. Conversely, 'statistical wrestling bouts' would be necessary to deliver the repercussions from their global anonymization.

6   Scientific calculation of risks remains locked in a circle of technological mastery. The abstractness of the calculation in relation to particular technologies guarantees that technologies can be compared; comparison with accepted technologies in turn guarantees calculability, which must for that very reason be unverifiably, circularly presupposed, compelling the denial of what can never be excluded: dogmatism.

7   The assumption of technological mastery turns into technological irrationality, in the face of hazards which cannot be technically excluded but only minimized.

8   The whole global experiment is calculated and evaluated from start to finish on the basis of probabilities, which allow the conversion of improbable counter-examples into confirmation; and the only 'significant' tests available for these would far outstrip both the lifespan and the imagination of human beings.

9   Research is obliged to enter into a pact with capital, wherein the research findings decide whether a profit or loss is made. The potential refutation of the original assumptions takes governments to the brink, threatens to set off an avalanche of compensation claims, and, not least, causes faith in the epistemic capacity of technological science to melt away.

10 All the divides between theory and practice are abolished: the lack of responsibility for consequences; the disinterested testing of theories; freedom of research without the obligation to apply its findings. These no longer determine the reality, only its representation.

Hazards are technologized away into discrete compartments; repercussions are statistically anonymized; the criteria for strict causality are strengthened; and the technological and natural sciences hold a monopoly on high-tech, expensive research facilities. Thus science and technology deceive themselves about the fact that they are, in the nuclear age, no longer an experimental, empirical science as their founding fathers understood the term. The theories of safety, which are supposed to minimize the hazards, have invalidated experimental logic. But falsified theories do not care a hoot, provided that their practices guarantee success, strengthen the backbone of external interests, and there are no alternatives to speak of.

None are so blind as those who will not see. In his autobiography, Arthur Koestler describes how he travelled in the Soviet Union in 1932 as a convinced communist, and saw only too well the human distress and indications of the Stalinist terror, but allowed his ideological training as a party official to explain these away. Have the natural sciences of today become, in respect of training and research, a kind of cadre school teaching one not to perceive the reality of the dangers produced by the system?

Yet there are surely deeper, simpler reasons for this ignorance. One can dispense with learning from one's mistakes. So too with religion (begging your pardon). If, however, one doubts whether technology and science will still be necessary tomorrow, one might as well doubt whether it will still be necessary to breathe tomorrow. Perhaps there will be a variety of alternative forms of science, of which we have as yet no conception, in the future of scientific-technological civilization, but not an alternative to science.

Human error is not the cause of all evil, as they try to tell us after every accident. Quite the reverse, it is an effective protection against technocratic visions of omnipotence. We humans make mistakes. That is perhaps the last certainty that remains for us. We are entitled to make mistakes. A development that forecloses this leads further into dogmatism or the abyss – and probably both.

It is accordingly of key significance that a critique be made possible at the centres of industrial-technological development. It is only where nuclear physics, medicine, information technology and human genetics take a stand against their respective domains that the world outside can see what future the test-tube holds in store, and what dangers. Critique and resistance in the technological professions improve everyone's chances of survival.

6

The 'Poisoned Cake': *Capital and Labour in Risk Society*

Chemistry consumes life. Every day the mass media find new things to report: of wine, cheese, oil, vegetables, furniture, paints, detergents, veal, and so forth. Only a few years ago everybody would still have believed, indeed proved, the pointlessness under capitalism of poisoning the consumer, who is supposed to swallow and pay for everything. Yet reality refutes even the most conclusive proofs.

Labour and capital, along with politics, the economy, and the social sciences, remain locked in the industrial harmony/opposition which has successfully equated economic upturn with social progress. Initially this only means that the cake to be divided up grows bigger, and that the conflicts underlying its production can be assuaged under constant or even increasing inequality. Even someone who supposes, with Marx, that the abolition of private ownership of the means of production takes us to the threshold of a better world, presupposes that the development of productive forces draws forth an abundance of wealth, and thus of conflict.

Where the cake is poisonous, however, this cosy industrial world of stabilized conflicts goes off the rails. And precisely therein lies an essential reason that the poisoned cake should be and remain non-toxic – at least, in the imagination of the key players.

It is like a good old music-hall turn. So long as the soup tastes good, one can agree to disagree about the size of each helping. But what if it is, or seems, poisoned? Here we have the first dispute over principles. Suddenly there are only opponents, who do not even know precisely whom and what they are against. Nobody owns up; nobody wants to finish the soup or to pay up. It was the others. Everybody did it, so nobody did it. Those who claim that it is poisonous are liars. Silence detoxifies, just as in real life.

Is this struggle to distribute costs away just an additional dispute, a secondary conflict, or does it change the logic, structure, mode of expression and predetermined loyalties of industrial conflict? Do new, independent arenas, instruments and power positions unique to this confrontation emerge with the struggle over the poisonous cake? Do old and new forms and lines of conflict overlap? Or do the axes of the conflict structure shear and interpenetrate, so that in the end a different social structure, a new historical entity arises?

None of these questions is answered here, even where more forthright answers are attempted, unhedged by qualifications. The answers only ever serve one purpose: as an underlying basis for the questioning. The latter is a matter of urgency, and must first be wrested free of the consensus of industrial elective enmities that distorts it.

Relations of production and relations of definition

The industrial world as it is represented is all about increased production and product margins, profit opportunities, market shares, income, patents, job security. These are partly the preconditions for, and partly the consequences of, survival in the struggle over the distribution of social wealth. Ownership – of capital, but also of professional qualifications – and the possibility of cornering or opening up markets, setting up or breaking monopolies, constitute the 'rules of the game', demarcating the horizon of possible action. All this has frequently been thought out and researched, though the questions have not thereby become any less clamorous or more answerable.

If (without coming to any premature theoretical conclusions) we attribute this to production relations, then the dispute over the cake's toxicity brings into focus other arenas, means and positions, summed up here as the relations of definition, that is, such questions as: by what rules is the toxicity of the cake judged? How are the burdens of proof distributed? Must the industrial manufacturers present the proof, or are they given the go-ahead provided that the injured parties cannot prove its toxicity? What rules of attribution are applied? When, that is, is the case against a 'culprit' considered proved? What role do standards of scientific proof play in the process? How are compensation claims regulated? Who is able, and how, to enlighten the public about concealed pollution in the face of parties with a clear interest in denying its existence?

It is easy to perceive that the relations of production and definition have several similarities: they are power relations, i.e. they establish opportunities for access and implementation; they are both directed at the dispute over the cake of 'social wealth'; the history of their emergence is a shared one; both are based on old inequalities, and were fully articulated in Europe under the industrial capitalism of the nineteenth and early twentieth centuries.

Yet there are also major differences. These remain concealed as long as the distribution and regulation of early industrial risks substantially follow the logic and conflicts of wealth production. Relations of definition are not property relations, but basic principles underlying industrial production, law, science, opportunities for the public and for policy. They are employed in deciding about toxicity, and consequently about the utility and exchange value of goods. Thus they are concerned with matters that, while eluding sense perception, nevertheless globally affect health practices and the structure of norms of the security state. Relations of definition thus decide about data, knowledge, proofs, culprits and compensation, thereby also determining whether hazards can be anonymized and covertly spread; and, conversely, whether these concealment practices can be counteracted with any prospect of recognition. All this takes place against the broad backdrop of the profound political and economic effects of acknowledged hazards. It is a question of rules for the acknowledgement of injuries, despoliation and threats; rules which by no means have to arrive at an unambiguous result, but can also lead to ramifications and into grey areas, where only one party concludes, with good reason, that acute pollution has taken place. The social definitions produced here are, accordingly, the unstable products of social confrontation against the background of prevalent rules of attribution, legal responsibility and compensation. These are as much predicated upon cultural perceptions as on the reigning standards of causality and principles of culpability, blown now in one direction, now in the other, by the winds of knowledge.

As a rule, current relations of definition arose in order to compensate the repercussions of industrial production, and to safeguard industrial power relations. Yet they can become independent of production relations. Certainly, the development of social wealth goes hand in hand with a reform of the social relations of production (in the sense of redistributions, trade-union organization, social safeguards, etc.), and in part also produces it. Yet the qualitative historical change in the hazards has meanwhile highlighted the inequality of the prevailing relations of definition. These guarantee normality and thereby render themselves absurd.

The hazards to which we are exposed stem from a different century than the safety pledges that attempt to contain them. Therein lies the ground for both the periodic eruptions of the contradictions of sophisticated safety bureaucracies, and the possibility of constantly renormalizing these 'hazard shocks'. Institutionalized blindness to hazard, the confusion of the centuries that technological-bureaucratic handling of hazards has become in the nuclear, chemical and genetic age, also permits the absorption of protest. The historical maladjustment is 'functional', in that it has turned its denials into a system. In recognizing the basic rules, protest against the despoliation of nature and of humankind turns into the agent of its own refutation.

Similar (and yet hardly comparable) to the development of the distribution of industrial wealth into class conflict directed against the dominant relations of property and distribution, the production of hazards exacerbates conflicts over the dispersal of costs, with the ultimate aim of changing the relations of definition: the redistribution of burdens of proof; a change in the social construction of responsibility, causality and guilt, which underwrite the long-term generalization and legalization of annihilation hazards (chapter 7).

A crucial issue in the conflicts over definitions is who is to 'prove' the harmfulness of the substances, emissions, etc., and to whom; what counts as 'sufficient proof'; and how action, proof procedures and compensation are related to one another. A pattern has emerged here, under the power relations of industrialism, which can be designated 'faith in progress'. It is, in a certain sense, a form of extremist optimism where the burdens, not only of the damage but also of proof are foisted on the injured parties.

The injured parties, i.e. (almost) everybody, can never complete the obstacle race of impossibilities. First, they must demonstrate the injury; second, they must give proof that a given substance was indeed what gave rise to the injury; third, they must prove that the substance originates in a named company; and, fourth, they must finally accomplish the task, impossible against the inertial mass of data, of suing one person for legal responsibility.

What has to be overcome everywhere is not merely individual resistance, but the concealed gap of the century: dangers worldwide make it harder to prove that a single substance is the cause; the international production of harmful substances works against proving the culpability of a single company or perpetrator; the individual character of criminal law contradicts the collective danger; and the global character of the danger has abolished 'causes', as our industrial forefathers understood them. So one is welcome to search for them. Any laws, ordinances, inspectorates that leave this truly extreme inequality of the burdens of proof untouched will be unable ultimately to break the current practice of legalized universal pollution.

> Our legislation on civil liability is designed for injuries of structurally simple origin (accidents, acts of violence), or for events involving clearly identifiable injuries, taking place among a small number of easily identifiable individuals. That means that both of the functions of this legislation, namely compensation for damages or injuries suffered, and the preventive function of behavioural control, are largely ineffective in the context of environmental protection. In respect of the principle of culpability, however, the behaviour-guiding effect of the law of liability would be of special significance. For it is a question of internalizing the possible consequences of an environmentally damaging action, thereby contributing to better treatment of the environment. In so far as the civil law does not exercise this behavioural control it also fails in its preventive purpose, which is supposed to determine the content of environmental policy and environmental law according to the principle of prior provision. This result is bound to make one uncomfortable. For, contrary to an elementary precept of distributive justice, the burden for environmental damages is frequently not borne by its perpetrators, but foisted, via the principle of burden-sharing, on the environmental victims or on the state. (Ritter 1987, p. 935; for an early study of these questions cf. also Schäfer 1972, pp. 125ff)

It is this system of organized non-liability that simultaneously lends a 'residual meaning' to all reference to 'environmental danger'. The 'environment' is stabilized not by nature, but by the rules that allow something to be used as an 'exterior', projected outwards – simply because the inequality of the burdens of proof and the impossibility of attribution allow this.

If the burdens of proof are redistributed, then what can be foisted on the external world (under the normal pollution sustained by current rules of anonymity) strikes at the centre of industrial management. If the economy is required to prove the claim of being responsible, its claim will hinge on proof of non-toxicity.

The talk of 'environmental hazards' is a protective claim, expressing a collectivity's defensive and wishful thinking that takes the hope for the reality, just as the *Titanic*'s crew lulled itself into the certainty that the ship's hull would manage to plough through the icebergs. It is the principle of 'not in my backyard' of a whole epoch yielding itself up to the illusion of living close to nature, where nature has long since been exhausted, deformed and thus destroyed in its very assimilation (see chapter 2).

What one sector, e.g. the chemical industry, contaminates *qua* environment is what others, e.g. tourism, agriculture, the food sector, fisheries, etc., supply to the market. Thus natural despoliations also turn, in an immanent contradiction between the interests of entrepreneurial capital and rentier capital groups, into product despoliations, the destruction of property and market sectors.

Thus considerable conflicts of interest within the economic camp are exacerbated, under the catchword of 'environment'. These interests are apparently separated from one another by the definition of environmental despoliation, but in fact they operate in the market system as a kind of Russian roulette.

The social struggle for acknowledgement of pollution's toxicity is thus also a confrontation within society: there is a new scenario with strategic positions and instruments, which must now be submitted to a social-scientific analysis.

Unattributability as a system: unpolluted because polluted

Where there is pollution, it is difficult to solve the riddle of responsibility. The distinctive feature of the theatre of poison, on whose stage we act out our lives, inheres in the fact that both culpability and innocence are provable under the prevailing rules of proof. This guarantees perpetual acquittal, as well as perpetual accusation. At the same time it casts suspicion of complicity on the means of proof. In the confusion that results, the good and the evil, plaintiffs and defendants, spectators and participants, all become increasingly difficult to tell apart. Therein lies the performance's fidelity to reality. But there is another, long-term effect: every acquittal is an accusation. On the one hand, acquittals are now the almost unbroken rule; on the other, almost everyone is now a defendant. Each conditions the other. Even if it never becomes quite clear whether everything or nothing has been polluted, it is clear that the climate is becoming polluted. Self-evident facts have gone on strike. To see how, one need only study the drama of reality itself.

The beginnings are usually alike. The outward appearance, smell, etc., is deceptive. Even where the pollution is obvious (lungs, trees, statuary, etc.) it cannot be spoken of . 'Reality' proves this according to its specific 'logic'. Let us take a quite normal case, which brings out the nature of the proof: the proceedings against a lead crystal factory in the district of Altenstadt in the Upper Palatinate.

Penny-sized flakes of lead and arsenic fell upon the area; wreaths of fluorine fumes turned the branches brown, etching windows and crumbling roof-tiles. Inhabitants suffered from rashes, nausea and headaches. The people who had to endure all this were in no doubt of its provenance. White dust billowed out of the factory chimney, and not only the grass was discoloured. For the sceptics, measurements confirmed that pollution was many times in excess of the normal pollution permitted by maximum safe levels. Moreover, further inquiries revealed that several of the company's plants were being run without a licence from the authorities. A clear-cut case.

A clear-cut case: on the tenth day of proceedings the presiding judge offered to stop the trial, against a fine of 10,000 DM. Such an outcome is the rule for environmental offences in Germany, as statistics confirm. The number of identified cases that go to trial has multiplied over the past five years (1980: 5,000 cases; 1985: 13,000 cases); in 1985 there were ultimately twenty-seven convictions with prison sentences, of which twenty-four were held over pending appeal. The remaining cases – just under 100 per cent, that is – were abandoned in spite of the flood of legislation, in spite of the public objections of ever more zealous and well-equipped inspectorates invested with police powers.

How is that possible? It is not only the lack of laws or the proverbial failure to prosecute that protects the culprits. The reasons lie deeper and cannot be overcome by a full-throated appeal for policing and jurisdiction, such as the environmentalists are increasingly making. It is in the strict application of the principle of culpability that everyone seeks salvation. This is supposed to lead to judgements, but in fact prevents them. While it is necessary to seek the 'culprits', the search loses its way in the 'universal culpability' that omnipresent danger has swollen into: the undeliverable proof turns into a preventive acquittal.

The density of legislation and of official controls, public, industrial and private expenditure on environmental protection – these are all on the increase, as are the recorded pollution levels and the lists of dying plant and animal species. The reason lies in the stability of the relations of definition, which emerged in the age of the individual culpability principle. In the present age of worldwide traffic in toxic and harmful substances, the relations of definition turn the legal system into an accomplice of ubiquitous pollution, which cannot in principle be proved to stem from an individual.

The culpability of the lead crystal factory was undeniable, and no one denied it; but, in mitigation, there were three more glass factories nearby producing the same filth, and these were also being prosecuted by the state. Note the consequence: the pollutants pumped out by everyone are pumped out by no one. The greater the pollution, the less the pollution. Whoever fails to see the internal logic of this 'more-is-less' relationship must take into account the principle of individual culpability. Ubiquitous pollution, blended with the principle of individual culpability, guarantees everything: acquittal, pollution, proof of non-pollution (which can be delivered via the impossibility of proof of having polluted) and proof of pollution, which, moreover, nobody denies, but simply remains unprovable.

For example, is the 'barking' of children in smog triggered by smog or by the weak constitution of the child? This is only one of the causal hurdles that the injured are capable of clearing only in the rarest of cases.

Suppose by some miracle the causal relation were demonstrable. The nuclear energy law would 'threaten' to fine the miscreant. A society in which anyone who pinches a radio from a caravan can be fined twice as much as an unscrupulous factory owner who permanently contaminates the entire surroundings with radioactive substances, cannot assume that the injunction to do what one can will counteract the devaluation of its legal principles.

If one wanted to think up a system for turning guilt into innocence, one could take this collaboration between justice, universal culpability, acquittal and pollution as one's model. Nothing criminal is happening here, nothing demonstrably criminal anyway. Its undemonstrability is guaranteed precisely by compliance with, and strict application of, the fundamental rule of justice – the principle of individual culpability, whereby both pollution and non-pollution, justice and screaming (coughing) injustice, are guaranteed.

More precisely, the trick lies in the fact that pollution is the reason for non-pollution: the greater the pollution, the greater the number of chimneys and waste pipes through which harmful substances and toxins are pumped, the lower the 'residual probability' that a culprit can be made liable for the collective sneezing and wheezing (which can also have quite natural causes), therefore the less the pollution. As a result (since the one does not exclude the other), the level of contamination and pollution rises.

The reality of industrial fatalism thus rests upon (at least) two simple cornerstones: the universalization of pollution and the principle of individual culpability. Its very realistic logic lays down that the greater the number producing it, the more obvious it becomes that not only this or that person produces it; hence it follows that nobody produces it.

In the no man's land between actual universal pollution and legal non-pollution, the legal system itself enters a twilight zone: what sense is there in a legal system which, in its democratic splendour and judicial independence, is condemned by its principles of decision-making to an idle, symbolic detoxification?[1]

The variability of the relations of definition

The frequently invoked 'natural destiny' of civilization is thus predicated on extremely unequal relations of definition. If, for example, factories had to prove what they assert – that they produce no more pollution than humans can bear – or be prosecuted, they would be deprived of the shield of unprovability. The reality would be the same. Only the burden of proof that the state of things is ideal, as they claim, would lie with the perpetrators. Hence it follows conversely that the remaining opportunities for foisting costs and pollutants on the environment are grounded in the system of organized non-liability. It is the radical inequality of burdens of proof, and the historical inappropriateness of the norms and institutions for attribution – and not only 'systemic differentiation' (functionalism) or 'the logic of capital' (Marxism) – that enable the companies to pollute the environment and saddle it with the costs.

While 'functional differentiation' and 'the capitalist realization process' have established themselves in the 'mechanics' of the social (which is nearly impenetrable to social action), the relations of definition are multivariable. Small things can have large effects. Take the example of California: some years ago a law came into force there to keep pollutants away from the population and, should this fail, at least to warn them about pollutants. Even more important, the burden of proof rests with the manufacturers of pollutants, i.e. with industry. This was approved in November 1986 by a two-thirds majority. The environmental protection given by the law differs from customary legislation in three important respects: first, carcinogenic substances or those harmful to the unborn may only be used in quantities representing 'no significant risk'; second, industry is obliged to take care not to expose the population to any significant risk, and must prove this if necessary; third, any citizen can enforce the law individually by bringing defendants to trial.

The consequences are considerable: now it is not only the oil refineries and nerve gas producers who have to warn that they have chemicals stored 'which, according to the information of the state of California, cause cancer, birth defects or other reproductive damage'; petrol stations, supermarkets and factory buildings are also obliged to hang up signs with similar warnings. The possibility is not to be excluded that companies, having learned their lesson from full warehouses and empty tills, will discover their most basic interests to be served by the preventive detoxification of their wares.

Capital against capital: the distributional dynamics of those who win and those who lose by risk

Whereas, in the old industrial society, the logic of the distribution of wealth coincides with that of risks, in industrial risk society these also diverge.

On the one hand, the rule continues to hold that wealth rises to the top while risks sink to the bottom. The attractive force that poverty continues to exert, as ever, upon dirt, destruction, toxins, noise, etc. – nationally too, but especially internationally – is not the reason for the epochal change. On the other hand, the industrial system is taking belated revenge upon those who have enjoyed its fruits until now. There is an essential difference between the battlefield of wealth production from which the nineteenth century derived the experience and premises of industrial and class society, and the battlefield of risk production in the developed nuclear and chemical age, to which we are only slowly becoming sociologically sensitive. This difference surely lies in the fact that – to put it simply – the production of wealth leads to the emergence of class contradictions between labour and capital, while nuclear, chemical and ecological systemic hazards lead also to the emergence of 'class contradictions' within the ranks of capital – and thus within the ranks of labour. The transverse differentiation of the social structure, the fact that the structure of industrial conflict melts and is recast in the heat of hazards, is the most menacing and inflammatory problem from the point of view of the social structure and the economy.

The social and welfare state had to be realized against the mass resistance of private investors, from whom the costs of increased wages and overheads were demanded; ecological dangers realign the economic camp. New links emerge between those who profit and those who lose by risk in the economic system, sometimes along industrial and sectoral lines, but sometimes also within plants, from one sector and product type to another. It is not at all easy to discern the boundaries at first glance, or, more precisely, to discern who obtains the power to draw the boundaries where, and on what grounds.

Only new concepts and distinctions can open this field to research. The initial insight is that under the heading of the 'environment', damage to markets is *de facto* passed on and anonymized. What chemical plants, etc., direct into rivers and seas is caught by fishermen, who supply it to their clients in the marketplace in order to survive. Yet pollution turns the conditions upside down. The poisoned fish catches the fisherman in his own ecological nets and, in so far as its toxic content scares off the consumer, drags him onto the dry land of insolvency.

Certainly, if some industries lose, others profit, making a mint (or rather, a pile of forged banknotes) out of the dangers. To this extent, the public drama of risks enacts a social displacement, a redistribution of profit opportunities (cf. Jänicke 1990).

Today, a low-growth economy is already beginning to discover in hazard creation the instrument, as it were, for developing and stabilizing a consumer-independent market. Hazards represent a kind of compulsive marketing.

Where the world has become perilous, the world population is becoming a consumer of hazard prevention. The more menacing, encompassing and ineluctable the hazard, the more inexhaustible is the market for 'fighting' it.

Further talk of 'need' and 'demand' in this context is superfluous. The fact that the individual can barely escape the hazards puts the automatic mechanism of compulsive turnover into gear. There is surely no historical precedent for this. One gets an inkling of the global market forces being unleashed here from the impact of the nuclear shock on the growth of the Geiger counter industry; or from the exponential growth in industrial pharmaceuticals, fostered by the worldwide hope of a cure for Aids.

The new 'industrial double income' – chemistry fights the repercussions of chemistry, medicine fights those of medicine, etc. – also belongs to the collection of curios that ought to be analysed by ecology and economics. Large chemical companies, for example, profit both from environmental pollution and from environmental protection. It is no wonder that they applaud protest through the proper channels, and it is precisely those companies which contribute most to the desertification of the environment that are to be found among the first sponsors of civil protest (cf. Enzensberger, as early as 1973).

So there is confusion about confusions, material enough for whole generations of researchers. The new and surprising fact, however, never sufficiently recognized or worked out theoretically or politically, is that the industrial system does not only 'profit from its iniquities' (Martin Jänicke), but that new alliances are formed precisely thus in the economic camp. In the 'environmentlessness' which the delicate society–nature blend we inhabit has become, 'environmental' despoliations always also signify the destruction of property and market shares. This relationship remains concealed in unattributability *qua* system, which is finally, loosely legitimated in talk of 'environmental' dangers.

In this sense it is bitterly ironic that industry should thank the environmental movement, which, for all its political volatility, ensures that the battlefront between the winners and losers of social risk situations remains concealed.

It may still be possible to speak of the 'environment' at the individual factory level, though this too is becoming more difficult, more tenuous as it were, given that high chimneys, for example, are becoming a politically problematic condition for internal cost calculations and managerial autonomy; talk of the environment at the overall social level is, however, simply fictitious. If it is suddenly revealed and published that certain products contain certain 'poisons' (information policy is becoming a key factor, in view of the usual imperceptibility of the hazards in everyday life), whole markets collapse, and capital investment and production are devalued at a stroke. This 'ecological expropriation' thus constitutes a historically unprecedented devaluation of capital and productivity under constant property relations, usually without any change for the consumer in the appearance or utility of the goods.

Product hazards and production hazards: the Achilles' heel of the manufacturers of danger

The distinction between production hazards and product hazards is central to this effect. The difference lies in the companies' opportunities for passing the buck, and for strategic anonymization under conditions of organized non-liability. Whereas production hazards can be foisted on the environment (at least from the company's point of view), the product is ultimately always identifiable with the company, which can never entirely reject claims that it is liable. Toxins, risks inherent in the product, counteract the constructions of anonymity. Thus palming off risks on the consumer becomes economically risky for the businesses themselves. They must also expect people to boycott their products, even where they are legally protected from liability claims.

Product risks always carry an Ariadne's thread within them leading out of the labyrinth of non-attributability. Unlike production hazards, which can be shrugged off because they are protected by the unequal burdens of proof, product hazards are subject to the 'boomerang effect': ecological expropriation (potentially) affects the ecological expropriator.

This sensitivity of the market to hazards grows with product availability, and with the sensitivity of sales to public perceptions: where the 'hazardousness' of an internationally marketed gas lighter is taken up by the mass media, even a giant company's market position can be severely weakened.

Old products and established markets are in this sense also 'susceptible to definitions of risk'. Change the legal or scientific rules, decisions, results, and you change the subject of cultural and mass-media attention; markets can collapse overnight, together with everything that depends on and lives off them. Turnover predicated on science ends up dependent on the variable conjunctures of presumed scientific knowledge. If, as tends to be the scientific norm, knowledge is a function of change for its own sake, and for the sake of individual reputations – if, for example, unleaded petrol is suddenly no longer 'environmentally friendly', but 'carcinogenic' (as has been reported in the media) – then the planned products, markets and policies built on presumed scientific knowledge collapse. In this case the catalytic converter industry is affected, which is predicated on progress towards a lead-free environment. This does not involve damage to the interests of a 'marginal industry', but repercussions against industrial power centres. Normality, i.e. the malleable, protean character of scientific findings jeopardizes the business interests that are based upon them.

The production of 'product risks' is the actual Achilles' heel of the risk producers. For these risks represent a threat to manufacturing industries, and to all the businesses and groups of workers that depend on them. This hazard increases with the liberalization of safety standards, which are supposed to remove it.

If one relaxes the limits on maximum pollution levels, one decreases the potential number of toxins that can be apprehended (and thereby officially acquitted of suspected toxicity). The grey area of legalized normal pollution grows, and the relevant markets are rendered more vulnerable through the public airing and discussion of suspicions in the mass media. Companies are sitting, so to speak, on the powder keg of anonymized universal pollution. One might say that the legalization and anonymization of the hazards returns, being redistributed in the form of increasing economic uncertainty in markets and industries that process or retail 'ecologically risky' products.

In the process there emerges a phase difference within the risk society, in respect of the systemic opportunities for the anonymization of hazards by their producers. The triumph of genetic engineering leads to a generalization of product risk. An entire future industry has put an economic noose around its own neck, and the increasing public awareness of hazards is tightening it.

To put it in historical perspective, while the nuclear power industry conjured up cost-intensively minimized manufacturing hazards of unassessable extent, industrial chemicals have become major production and product risks. The dawning genetic age, however, is backhandedly reintroducing industrial product liability, and thereby undermining organized non-liability. When something happens, the industrial profit-makers will this time be sitting in the trap of individual culpability. Perhaps a realistic estimate of the hazards and market perceptions is already inhibiting the 'genetic gold rush'.

The case of production hazards is different. 'Environmental' destruction, whose unprovability is predetermined, volatilizes whole entrepreneurial sectors and industries. The economic destruction of the human environment appears like a bolt from the blue, with the mass media publicly revealing and denouncing the relevant product pollution and contamination. The mass-media spark ignites the flame of information policy – and people are affected who neither play any part in the pollution nor are able to counteract the pollution of the wares that they offer. The fishermen, farmers, holiday resorts, etc., industries that survive by the commodification of contaminated 'nature', become the first 'enterpreneurial proletariat' of ecologically reckless industrial production.

Yet there are no fixed, secure boundaries between those who profit and those who lose by risk. It can strike at a great number of the players, if not all, as the shifting foundations of knowledge dictate. Yet that means that public awareness of the dangers, with the participation of many institutions and groups – in research and television, law and policy – undermines the economy's autonomy, drawing the economic system into social disputes, down to the details of its production. The universalization and suppression of the hazards now conjures up the dread spectre, hitherto kept at bay, of economic planning: public, social involvement in and control over companies' decision-making processes.

The politics of definitions: the fragility of markets in risk society

If the 'latest' researches reassess the toxity of a given foodstuff, then the latter may be transformed into life-endangering stuff. There are surprising upturns in alternative markets, and floods of compensation claims are triggered off. Different rules for causal attribution and jurisdiction for what is constitutive of proof, for establishing maximum pollution levels: these redistribute the winners and losers, and bring about new economic alliances.

The systemically manufactured and denied self-jeopardization of the economy (which is bound up with the disintegration of prices, markets and capital) haunts the economic system as a kind of 'destructive anarchy', whose explosion is touched off by the socially manufactured 'spark of knowledge'. The other side of the coin of the repressed production of danger is a new friability of markets and of profit opportunities in the risk society. The latter's markets are built upon the card-houses of relations of definition, and these can be knocked down merely by the wind in the mass media or by changes in public perceptions. Capital exploitation becomes more dependent than ever on public opinion, research and culture.

It may be that 'everyone is in the same boat' in the flood of hazard. But as is often the case here are captains, passengers, helmsmen, engineers and people drowning. Some industries cannot profit from dangers, and are only vulnerable to them; others combine great 'flexibility to dangers' with the chance of transforming them into opportunities for introducing new products. In hazard-conscious civilization, whatever provides for, improves or prevents 'flexibility to dangers' in products, markets and companies becomes a matter of economic survival.

Also, the chances for taking part in the negotiation and shaping of situations of social danger – that is, where to locate and assign 'causes' and 'culprits', which counteractive measures to take and which to exclude, and so forth – are in fact very unequally distributed in society. Industries that only live off what others contaminate as the environment, having no lobby at their disposal and staking their economic survival on increasing public knowledge of the dangers, oppose others which produce toxifying and detoxifying technologies in one and the same enterprise. These have manifold opportunities for slipping past external controls by using their monopoly on information, as well as influencing the process of defining risks through public relations and lobbying in the political arena.

Every time a dangerous situation created by industry is recognized, e.g. the dying forests, the question of its 'causes' and 'culprits' is put on the political agenda. Yet which of the possible partial causes are truly social, a lever for diminishing hazards, is not pre-decided. Where the 'spotlight on causes' sheds its light, 'causes' turn into 'culprits'. The struggle over percentages becomes a displaced struggle for markets and power. In the case of the 'dying forests' (cf. Roqueplo 1986), the scapegoating of cars was the result of long 'definitional struggles' fought out with scientific evidence. As a result, private traffic meant that the population itself was the party mainly responsible for the death of the forests, although only 30–40 per cent of the pollution through nitric oxides is attributable to cars, while industry accounts for the lion's share. This not only provided a protective screen for industry against external intervention; in a sense, the population was itself to blame for the destruction it protested against. The car industry, however, after initially resisting its good fortune, was able to maximize its profits with 'environmentally friendly' cars.

The 'definitional struggle' does not begin with the drawing of (practical) conclusions, but in the act of definition itself. Definitional vagueness is also and substantially socially conditioned, an expression of the fact that 'any situation pregnant with risk – particularly if the risk is a priori of human origin – fragments the discourse into antagonistic, contrary discourses' (Roqueplo 1986, p. 404).

Strictly speaking the expression 'acid rain' refers to a rising acid content in the air, caused by humans. The main culprit is sulphur dioxide (SO_2), which is a product of combustion of fossil fuels (coal, petrol) in private homes or in industry, in conventional power plants and heavy industry (smelting, cement works, coking plants, etc.). As for the ecological consequences, these include not only dying forests, but also dying lakes (Canada, Scandinavian countries) and damage to sites of cultural importance (corrosion).

The term 'air contamination' (or 'pollution') has a different signification: it connotes not so much forests and lakes (which are located as a rule in regions of low air pollution) as health problems arising in heavily contaminated areas. This also explains why instruments for measuring air contamination levels are almost exclusively set up in these, for the most part urban, regions.

The expression 'dying forests' is ecologically more to the point, but it refers to an ensemble of causes not reducible only to air pollution. Yet by increasingly taking air pollution to be a primary factor in the death of forests, one tends also to give extra stress to the role of photo-oxides over acidity. Now photo-oxides arise from unburned carbohydrates and nitric oxides, whose increased dispersal in the air is a direct result of the growing supply of cars. Thus the connotation is clearly different here: it is a matter of the detoxification of exhaust gases, of catalytic converters, unleaded petrol. (ibid., p. 403)

Considerable social, political and economic investments are at stake, whatever definition one chooses. If one defines 'present harm to the forests from air contamination, caused in particular by smokestacks and the automobile industry', then one puts fossil energy production and the automobile industry in the dock. If one insists that definitions are vague and research incomplete, then the economy's potential 'defendant' sectors are let off, but the poisoning continues. From the French point of view, on the other hand, what is involved is a 'collective psychosis transformed into profits by the German automobile industry, which – while threatening the survival of the European Economic Community – allows the establishment of regulations that help it to dominate French, Italian or British competitors. (ibid., p. 404)

In the search for a remedy one can, and indeed must, always allocate different weights to various causal factors in the debate about the 'dying forests', in each case with very different in-depth political and economic consequences. Yet wherever a halt is called to the causal regression analysis (in principle inexhaustible), employment-creating firms and industries turn into 'suspects' and 'culprits' in the social domain. All this – the necessity of selection; the politico-economic 'valency' of the causes; the law that the cut-off point for causal explanation politically designates the point of intervention – turns natural-scientific theories and data in social conflicts over risk definitions into the playthings of politico-economic interests – and into the subject of sociological analyses. That does not mean that one can proceed at whim in public 'battles over definitions'. 'Hard' natural-scientific results are and remain the medium of the confrontation, and of the justification for the moves in the game. But natural-scientific complexity cannot be overcome by natural-scientific means; it remains, rather, subject to social and political choices. Thus two planes interpenetrate and enclose one another in the social definitions of risk: outwardly it is a question of natural-scientific findings and causal interpretations, yet political and politico-economic trends must be set thereby. The 'scapegoat' function of the car vividly demonstrates this.

> *Question*: Why did the issue thrown up by the dying forests concentrate on the car?
> *Answer*: The population was represented as being itself responsible for the dying forests, through the car. Now the car only accounts for 30–40 per cent. The rest is caused by industry.
> *Question*: Might there perhaps be a taboo against speaking about industry?
> *Answer*: Yes, of course! Zimmermann's [1985] strategy consists in ascribing a problem whose chief cause is industry (above all, power stations) to individual behaviour, and thus concealing the problem by saying, 'You yourselves are to blame.' That is like taking people to task for discarding sweet wrappers and saying, 'Don't talk about industry while you yourselves display such culpable behaviour.' (Roqueplo 1986, p. 416)

Where everything is polluted, disclosure or concealment of a given example determines whether markets collapse. This is equivalent to an empowerment of information sub-policies in the mass media, in research and administrative control institutes and in the law. A 'purely factual' TV programme shown at prime time has approximately the effect of a parliamentary decree, together with its enforcement, banning an industrial sector. Control of scientifically established information about dangers, and of the channels for its dissemination, becomes of crucial importance in the definitional struggle over the assignment of dangers and market shares. This power displacement takes place, moreover, under no formal change in the power structures. Employees in the scientific and mass-media institutions that exercise power through information do the most legitimate thing in the world, pointing out health hazards. Thereby they achieve power over power, without exercising power in the classical sense.

As cultural sensibility increases, the social definition becomes increasingly independent of natural-scientific diagnoses. A danger is economically real if it is socially perceived as a danger. Thus the power of the scientific, official and mass-media control and information institutions grows with the universal dissemination of poisons on the one hand, and with cultural sensitivity on the other.

The enterprises, companies, communities, industries and proprietors currently or potentially struck by this 'blow of fate' find themselves in an extreme situation. The publicly enacted devaluation threatens their economic existence irrespective of their performance, and frequently through no fault of their own. Thus the contradiction between the immanent production of systemic hazards and their non-attributability turns into the contradiction of an arbitrary, anarchic jeopardization of property, performance and capital. It is an entirely open question how this dependence of whole industries on the definition of risk, as a new dimension of the struggle for survival, will be worked out psychologically and politically by those affected, both in the short and in the long term.

Certainly, those who profit by risk are still preponderant. The most endangered are still 'marginal groups'. The constructions of anonymity and of latency are still holding. The losses and destruction are still being normalized as 'fateful blows' from competitors. The blind rage of those threatened is still being directed predominantly against the informants, but not against the chief perpetrators and legitimators of the hazards. The injured parties still have less interest than any others in complaining of their injuries. But what if 'ecological expropriation' becomes more than a mere threat to whole sectors of the economy?

What will happen after a nuclear, chemical or genetic disaster in which whole countries are (or are considered to be) contaminated, and the survivors find themselves collectively stigmatized as 'industrial pariahs'? Will there be ecological mass uprisings in the wake of a breakdown of the social institutions of control and legality, compared to which the slave rebellions of civilized antiquity were only a rustling in the leaves of the social order?

Even though this is not (yet) happening, it remains true that the minimization of conflict through economic growth is undermined by the intensification of conflict through the growth of dangers.

Labour society as risk society: the destiny of workers 'at third hand'

There would appear to be obvious conclusions to be drawn concerning the situation for employees and for union power, as it is affected by the polarizations between winner and loser industries in the game of 'ecological roulette'. The polarizations cut across the structure of wage labour. The lines of conflict in wage labour – in so far as the repercussions of industrial hazards and devastation are moved into the centre of the picture – could then no longer be attached to the criterion of non-ownership of the means of production, but to whether one's job is in parts of the economic system that profit from risk or in those that are endangered by it.

Hence it follows, first, that new kinds of contradictions emerge within industries, sectors, companies, i.e. between groups of employees, and therefore within and between union organizations; second, that these are contradictions at third hand, as it were, derived from the antitheses between fractions of capital, causing the 'fate of the working class' to become 'fateful' in a new and significant sense; and third, that the intensification and coming to consciousness of the corresponding lines of conflict can lead to an industry-specific alliance of the traditional 'class enemies', capital and labour, and consequently to a confrontation of this union–enterprise bloc with other partial coalitions, bypassing the entrenched class antagonisms which are forced together under pressure of 'ecological politicization'.

An example: after a television programme screened in August 1987 about worm larvae in fish, the fishing industry reported up to 50 per cent less turnover. Many small and family businesses were unable to hold out through the 'dry patch' and had to close down. There was more than the mere threat of job losses in the fishing industry, which employs around 12,000 people in the German coastal regions. Many businesses worked short time and had to dismiss employees. Apart from the job losses, this case concretely demonstrates the increased instability of markets, including labour markets, in a sensitized consumer culture. Moreover the result here is the same: the employees close ranks with their employers, and protest against the 'untrue' and 'exaggerated' representations in the mass media.

Now a case like this appears to give unshakeable confirmation to the honourable attitude that can justly be described as 'dominant'. The initial assumption in the public domain, in research and in the unions, is – to put it simply – that work experience and workers' experience and the desire to stay in employment prevent industrial workers from becoming ecologically sensitized, especially in the hazard-intensive manufacturing industries.

This antithesis between skilled workers and ecological consciousness of hazards appears to be an established fact of social science. In international research into attitudes on environmental issues, the rule of thumb is that 'green' consciousness comes with higher education; frequently the mocking rider is added that it is at its core a 'critical, teacher's consciousness' – which is intended to imply both 'generation of '68' and remoteness from technology. Sweeping statements to the contrary are often heard among the ecologically sensitized public, but also found in sociological literature: here the technological-scientific mentality is judged to be largely immune to learning from the repercussions of its mistakes, because of the specific character of its intervention into nature and humankind. Fietkau (1984) holds, on the basis of mass questionnaires, that there is a significantly lower behaviourally relevant environmental consciousness among industrial workers, corresponding to their instrumental view of nature, as compared with other population and occupational groups (admittedly this was before Chernobyl, Sandoz, etc.). In their own ranks, and in trade-union and related literature, attention is drawn to the field of tension between 'the environment and work'. Yet the prevalent conception is essentially one of 'rifts' through to 'contradictions', if only because the worker's consciousness, according to the traditional interpretation of Marxist theory, centres around the experience of labour, while ecological consciousness of hazards is sparked off by repercussions outside the factory. These special features and 'blinkered attitudes', as well as the elementary fact that knowledge of ecological hazards undermines one's own professional existence, are universally considered to prevent the cultivation of ecological consciousness and action at the heart of industrial labour, i.e. among skilled workers.

An investigation by the Sociological Research Institute in Göttingen, of which Heine and Mautz (1988) have published some preliminary results, thoroughly debunks this politically significant stereotype. Of the three occupational groups – skilled workers in the chemical industry (whose procedures and products are both subject to criticism), the unemployed and workers in ecologically neutral industries – they only encountered the rejection of 'ecological discourse' in 7 per cent of cases. Furthermore, as emerged upon closer analysis,

> even among those whom we class among the 'strict repudiators', isolated elements of ecological discourse suddenly break through the consistency of their argument, but even among the most indifferent (of whom there are astonishingly few), islands of ecological concern suddenly rise to the surface. It appears that ecological discourse has become a social reality unavoidable even for skilled industrial workers, albeit partly behind their own backs and against their intentions; we did not find any pure ecological nihilists among them.

To their own astonishment, and against their initial surmise, the authors encountered a marked ecological sensibility even among skilled workers in the 'ecologically suspect' chemical industry.

Barely half of the skilled workers at the chemical works noticeably re-
tracted or distorted ecological discourse on chemical matters. The contrary
was true of almost as many . . . i.e. their attitudes to their workplaces are
markedly more ecologically motivated than their behaviour at home, their
attitudes to cars and to the industrial future of the region. This result does
not correspond to the picture of a general tendency among workers in the
chemical sector to adopt a partisan approach to ecological matters where
their own jobs (in the widest sense) are on the line. Which ultimately
means . . . that the reception and interpretation of the environmental debate is substantially independent of, or even cuts across, sectional social
interests.

In their search for an explanation of this result, Heine and Mautz
discuss the thesis that the ecological sensitization of skilled workers in
the chemical industry takes place not in opposition to the factory, but
within it, in confronting hazards in the workplace, in expert consciousness and not least as a result of the ambivalent interests of
management.

Thus the management of the plant is walking a tightrope. On the one hand,
the potential risk arising from its own production should not exactly be
'overstated', in view of the fears of the population in the plant's vicinity,
and also in view of ecological suspicions; on the other hand, a working
attitude should be encouraged, among the employees at its own plant, that
at least supposes the existence of considerable potential risk. This second
consideration has long since created its own internal institutions and
didactic methods. In every factory, there are not only regular alarm drills
(which simulate a 'disturbance'), but also so-called 'safety demonstrations'
in which the risks of chemical production and of chemical substances are
demonstrated, sometimes in a drastically original manner: at one factory it
was reported to us that a lye used in production there was dripped onto the
eye of a freshly slaughtered pig, in front of the assembled shift workers, so
as to demonstrate 'how quickly it turns dull'. A succession of our inter-
viewees concluded, from experiences and instruction of this kind, with the
general statement that they had only learned 'how dangerous chemicals
are' through their work in industrial chemicals; some made a point of
mentioning that they had translated this knowledge from the factory to
their homes, attempting – as they said – to convince their wives not to use
such strong detergents any more. (ibid., p. 177)

While it was always problematic to sustain the widespread suppo-
sition that the management and workforce conspire to downplay the
risks, these results make it untenable, if only because of the contradictory cultural and economic interests on both sides. The modes of
perception and interpretation, particularly in the hazard-producing

industries, are split. This is no accident, and not (only) a matter of personality, but is predicated on systematic contradictions between safety claims and hazard production; at one time the costs demand to be cut and downplayed, while at another they underline the suscepti- bility of whole markets to changes of image and to public opinion. These internal splits are also of central importance to a counter-policy (see chapter 7 on this subject).

Yet the 'metamorphosis effect' continues to apply: ecological jeopardization endangers market strength and jobs precisely where the hazards increase while remaining concealed. One must keep in mind the possible significance of this for the labour and trade-union movements. The production and definition of danger relates largely to the product, which, almost entirely eluding the influence of the works committees and employee groups, falls completely within the sovereignty of management. This is, however, at the factory level. Though dangers are produced in factories, they are socially defined and evaluated in the mass media, in disputes between experts, in the maze of interpretations and liabilities, in lawcourts and in strategic, intellectual evasions – that is, in milieux and contexts entirely alien to the majority of workers. It is a matter of 'scientific struggles' above the heads of the workers, waged with intellectual strategies, in intel- lectual milieux. As things stand, the definition of hazards eludes intervention from workers, and largely even from unions. These are not even the primary injured parties, those being the companies and their managements. But as secondary injured parties, they have to pay with their jobs, if it comes to the worst, for the public consumption of definitions of risk.

Also, the definition of latent danger wounds their pride in per- formance, their promise of usefulness. Labour and labour power can no longer be conceived only as a source of wealth, but also, according to the widespread social perception of them, as an engine of threats and destruction. The work society not only loses its labour, and with it the only thing that gives life in it meaning and backbone, as Hannah Arendt ironically puts it, but also endangers this residual meaning.

If the management can switch to other products, then the 'ecologi- cal doubt' cast upon industries and occupational groups is exacer- bated for employee groups, jeopardizing their competence, devaluing work skills; for the latter, unlike products, cannot be exchanged. Nor does this happen *en bloc*, uniformly over all sectors, but it is split across the new, very mobile watershed between winning and losing sectors. Should this diagnosis be confirmed in the future, there will doubtless be considerable turbulence in the employee and trade-union camp.

This is because the switch to an ecological theme leads to profound shifts in and threats to the power of labour and its organized interests. First, the labour movement becomes doomed by what, in an earlier historical development, was a necessary precondition for its social existence: the decoupling of the product, the legendary 'indifference' to which the system compels and habituates waged labour, the ousting by the labour contract of the question of the social purpose and usefulness of the product of labour, and thus of one's own labour power. The labour movement cannot win the game of defining hazards, if only because it has itself withdrawn from it (out of sheer necessity) – legally, in the way participatory rights have been legislated and are practised; and culturally, in the blanket utilitarian assumption that the 'point' of an activity is proved by the receipt of wages.

Second, the loss of power consists in the splits encouraged in the labour movement by the effects of ecological devastation. The contradictions of capital precipitate out as contradictions of labour, objectively at least in so far as the labour market is concerned. That causes the fragmentation of the working classes, already foreshadowed after the advent of the welfare state by the weakening of class character and experience, to be exacerbated to an unprecedented degree (cf. Beck 1992, chapter 3; also Kern and Schumann 1984).

Readers are invited to fill in for themselves the details of this horrific scenario for the future of the unions. Let me add one practical political observation: if the labour movement does not intend to be elbowed out of history – and one would expect the future development of democracy to depend substantially upon that – then one ought to start in good time to formulate and implement an offensive, sociopolitically orientated 'product policy' in the context of the foregoing considerations; to claim and fight for a right of 'participation in product policy', both within the factory and at the overall societal level. The unions would have to denounce and dismantle the extreme inequality of the 'relations of definition'; create outlets for a pragmatic ecological critique, within each factory, of the latter's underlying rationale; fight for the right to technical criticism of the company, based on on-site experience and the 'perpetrator's' consciousness of responsibility. In other words, workers should not merely respond 'acidly' to ecological and other critiques, but accept them actively, preventively – perhaps putting themselves into unconventional coalitions at the vanguard of criticism; in particular, they should internalize the 'ecological capacity to learn', and foster and implement it in society. Such a 'greening' of the trade-union movement might well bring with it a new political spring. If Heine and Mautz's results hold generally, this can be achieved together with, and not necessarily against, an ecologically sensitized industrial workforce.

Regional strife: danger as internal and international conflict

It has so far been argued that ecological despoliation subverts the co-ordinates of industrial conflict in two respects: first, the environment is not only environment, but the economic basis for the existence of other branches of industry; second, environmental devastation foments polarization between winners and losers in the economic camp, and thereby also between occupational groups and their organizations. Next – and this step is intended to conclude the social-structural side of my argument – ecological despoliations lead to a specific type of conflict situation in (world) society: (a) these are no respecters of national borders, and thus displace the structure of interests within and between nation-states, as well as economic and military systems of alliance; (b) they are mediated by nature, i.e. the types of conflict they create, contrary to their own imperceptibility and abstractness, are perfectly concrete, in fact regionally identifiable; and these supranational, naturally mediated, socio-economic conflict situations are not 'acquired', but (c) geographically 'assigned'. Thus they stand in open contradiction to society's regulative principles of performance, property, etc., and therefore cannot in the long run be rendered anonymous by capitalist competition for markets, even under the cloak of normalized destruction.

In the highly industrialized world, the despoliation of nature leads the social schematism of wealth distribution – call it the 'class struggle' for simplicity's sake – to be freighted with and undercut by a vast regional conflict pattern, the regional strife of risk civilization. As conflict types, 'class struggle' and regional strife are essentially distinguishable by the latter's tending to arise from irreversible, naturally mediated geographical situations, the creation of 'toxin-swallowing regions', whose destinies coincide with the industrial despoliation of nature. Conflicts over risk definitions can be denied for a long time, but then precipitate into supranational, geopolitical conflicts. These can neither be defused through redistribution, nor justified in the guise of the familiar with the customary principles and consequences of competition and performance. They submit the nation-state's institutions for realizing the democratic will to an internal test of tensile strength.

If the North Sea dies or is socially perceived as a 'health hazard' – this distinction is immaterial to the economic effects – then it not only dies together with the life it contains and makes possible, but it also extinguishes economic life in all places, industries, countries that survive directly or indirectly from commodifying the North Sea. On the summits of the future that are visible on the horizon of the present day, industrial civilization metamorphoses into a kind of 'geographical (or 'world') estate' society, in which despoliations of nature coincide with those of markets. One's social standing and future are not decided by what one has or can do, but by where one lives and what one lives off; and by the extent to which others may use and pollute, with pre-determined non-attributability, one's possessions and skills as 'environment'.

Thus there is a limit to the fervent denials, whose official support is assured: the revenge of the abstract definition of danger is its geographical concretization. Every proposition can be brushed aside, the rose-tinting machinery of the authorities put into top gear. That does not prevent the destruction, but accelerates it. Yet the gravity of ecological-economic hazards, as against the authority of their denial, is based on their special political quality: on the devastation of socially internalized nature, the naturally mediated, creeping jeopardization of society. This takes its long-term, irreversible course behind walls of denial, on a material base of industrial, active substances.

A mood of panic reigns on Italy's Adriatic coast. Tens of thousands of tourists are taking flight. The hoteliers take one glance at their beds and beaches, which are emptying in spite of the high season, and the sunshine – and curse all the responsible authorities, as well as those not responsible. The whole Adriatic stinks to the heavens. For years in this country, environmental protection has been little more than a topic of conversation. Now rotting algae bloom is floating south along the Adriatic, in mayonnaise- and rust-coloured strands a kilometre wide. The River Po is taking its revenge on the tourist industry, having been contaminated by the effluent from thousands of sewers all over northern Italy, by fertilizers, insecticides, detergents and industrial toxins. Italy's largest river pours more than 47 billion cubic metres of chemically fertilized effluent annually into the Adriatic. Algae flourish there – particularly if it is as hot as it has been this year. The fish and the scampi, to which especially the German visitors used to be so partial, are suffocating by the ton from lack of oxygen. So the North Sea isn't alone in dying; the murder of the Adriatic is coming along nicely. (*Süddeutsche Zeitung*, 22 August 1988)

To put it schematically: class struggles are founded on production relations and take place between opposing groups within national arenas. In regional strife, however, groups confront each other whose social characteristics and shared traits are not based on their position in the social hierarchy; cutting across the latter, they are grounded in the geographical location of an industrially consumed and despoiled 'nature'. The contamination of marketed nature means that the regional situation coincides with the social one; more precisely, the former conditions the latter. Hierarchical inequalities, stratifications and class structures are freighted with or undermined by 'horizontal disparities' (Claus Offe), depending on river currents and wind directions, relative location of the sea, proximity of filtration and nuclear power plants, etc. Policy follows the fundamental principle: protect the toxin, in case of doubt, from the human intervention that endangers it; with that, geographical-ecological decay turns into economic-social decay. Supranational groups of regions and countries swallow poisons and waste on others' behalf.

Regions swallow not only the poison but also its non-attributability. Suing for damages helps just as little as protesting publicly. Whole regions end up in a catch-22 situation: they must demonstrate, yet may not do so if they are to survive. For on top of everything else, the 'poison-swallowing regions' are under compulsion to hush it up. Creeping, irreversible damage and subsidized, compulsive mendacity condition one another.

While the seal pups used in advertisements die wretchedly on the shore, it no longer helps to print glossy brochures filled with deceptive blue skies. One must take a realistic turn, and begin to awaken 'understanding', perhaps even putting oneself in the vanguard of Greenpeace, and at least using one's consumer power to declare oneself a 'chemical-free zone' (as has partly happened in the North Sea islands). Setting high maximum pollution levels only gives relief to the perpetrator regions, while exacerbating the dangerous situation of the loser regions. It seems a bitter twist of picture-book dialectics that hazards that elude our perception, and offer scant resistance to the manoeuvrings of definition policy, should become materially-regionally concrete; the more markedly and irreversibly with the increasingly successful staging and practising of their denial: the future of such downplaying of hazards shows in white spots on the map.

As has been said, industrial and ecological class conflicts and regional conflicts overlap. In the opposition and interplay between industrial and risk society, novel fusions and displacements of themes and areas of conflict take shape. For example, the battle in Germany against dying forests is evaluated and countered by its French neighbours as an adroit move by the German car industry against its European competitors (cf. Roqueplo 1986, pp. 419ff). In the end, no one is sure any longer whether the naivety of ecological zeal is as unalloyed as it gives itself out to be, or whether it is also or predominantly a 'dummy' strategy, converting world markets to the new religion of ecology in the name of a crusade for nature.

The 'industrial robber baronage' of the highly industrialized countries is well established, while the 'ecological crusade', the conversion and conquest of world markets, is only in its infancy. But it is already perceivable that a certificate of ecological purity can be exploited in order to sanction national prejudices and discriminations. This was discernible after Chernobyl, from the way in which the 'fraternity of socialist countries' was put into a kind of clan custody, and 'communist' lettuces, etc. were sacrificed to German fears; while in doubtful cases, western exporters were granted a rise in maximum pollution levels.

In these crusades yet to be unleashed on world markets, the (not quite) former colonies will end up in danger of an ecological recolonization. Not only do they offer cheap and (due to their poverty, as the cynic will affirm) 'voluntary' transit and final dump sites for the wealth of toxic/nuclear waste from highly developed 'muckslinging' countries. The market barriers, raised to new heights on ecological considerations, will put wall on top of wall, setting the seal on their poverty. Dangers and national policies will globally exacerbate the oppositions between the rich and the poor. Whole groups of countries will end up in the ghetto of the world's poorhouses, under the direction of international 'aid organizations'.

The regional and supranational chasms torn open by ecological-economic destruction overburden the political institutions of industrial society as conceived and organized in nation-states: first, they provide new networks between groups of regions and countries. A polarization emerges between the 'donor' and the 'recipient' regions of industrial waste, both within and between neighbouring states. Characteristically, it is not merely whole nations that are affected, but entire regions that straddle the divides of military alliance systems, ending up as it were in the role of the rabbits staring at the industrial snake. The regional 'poison-swallowing sites' that emerge, however, also straddle the policies of nation-states, the structured organization of their interests being neither provided for nor representable. The social structures of the production of hazards and of their consumption are different in kind, and belong to different centuries. The national-industrial decision-making structures of the production of hazards enter into contradiction with the supranational geographical sites where the hazards are consumed.

Yet regional conflicts, unlike class conflicts, have no nation-state as an arena for their enactment. They can dribble away, as it were, between the borders; or they can, through a geopolitical landslide, shift, abolish, undermine the borders and compel, or perhaps facilitate, a revision of the political map from below. For undelimitable hazard situations also have a pacific function, as yet politically underdeveloped, in producing shared features over and above the divides between military bloc formations.

Thus supranationalism both enhances the social explosiveness of the ecological crisis and encourages fatalism. In the end, national politicians can maximize their demands and go unpunished, as in the paragon of Europe, with its alibi of hyperactive inactivity.

Second, the legitimations of capitalist competition break down. One cannot hold bad management, underinvestment, overproduction, etc. responsible for economic collapse; the latter is a total process, transcending individual enterprises and industries, and solely or primarily due to geographical location in the interplay between nature and society. This assignment and identification of regions, however, stands in contradiction to the legitimating principles of capitalist competition for survival. There are certainly other, specific reasons for the infirmity of the tourist industry and agriculture: maximization of tourist capacity has led to the reckless deforestation of mountain slopes to provide ski-lifts and clear runs, transforming them into obstruction-free slides for avalanches. In the international structure of coercion, agriculture has turned into the site where chemicals are converted into life-endangering foodstuffs. Love for animals and a taste for them have combined in the consumer into a strange, defensive alliance. The normal person, partial to a bit of veal, is demanding more 'humane' methods of livestock farming, so that the little animals will taste better once they have been turned into escalopes and sausages. Yet these retrospective emendations of one's own 'destiny' cannot ultimately conceal that anything which flouts the fundamental principles of capital, property and performance, in the long term and with economic repercussions, cannot ultimately be integrated and legitimated by means of the political system. Lines of conflict emerge whose scale and unmanageability overstretch the institutions of nation-state, free-market democracies.

The new character of the situation certainly does not mean that a dawning awareness of the ecological powder-keg we are sitting on, or in, causes a great outbreak of unity: capital and labour as partners; red, black and green politicians celebrating a shotgun wedding; military blocs fusing in mutual trust; and people of various languages and cultures now giving common utterance to the distress that had silenced them. Yet, in the third place, the fact that whole regions are affected provides excellent opportunities for political resistance. Causal factors and effects are politically isolated from one another. The fishermen can hardly dam up the rivers of poison with their nets. Loser regions are thus politically disconnected from the causes and perpetrators of their destiny, and, by virtue of the political systems, only find very narrow channels for the representation of their interests. They have to shoulder the full burden of the discrepancies between official definitions of normality and the social perception of hazards. Gesture politics – the rain dance of critical loads, the talking drums of causal anonymization, the peace pipes of fictitious legal practice – all this adds public mockery to the derisory laughter of the hazards themselves.

Ecological-social 'regional conflicts' undoubtedly offer grand opportunities for suppression and concealment, but also exacerbate and inflame zones of conflict whose effects have not been comprehensible until now. Ultimately, the ecologically induced decay of whole regional economies coincides with the organizability of interests that can hardly be absorbed into prevailing political systems. Let it not be said (in an undertone of political realism) that this involves only lettuces, fish and holiday beds. The development of a leisure society is giving importance to recreation areas, both in the economy and in the consciousness of the population. Moreover, conflict situations are also substantially measured by their (non-)integrability into the established political system. In any event, localizable, socio-ecological lines of conflict – which need first to be lured away from their internal rule of silence and their industrial anonymity – intensify the regional contradictions anyway igniting within and between nations.

7

Conflicts over Progress: *The Technocratic Challenge to Democracy*

Environment as social habitus: interim assessment

This book is committed to the effort ('attempt' would be too weak a term, 'mission' too dramatic) to counteract the bewitchment of issues concerning society, culture and democratic decision-making by questions about technology and nature. If one pays attention, one can hear, growing louder and louder, the question: how should we live? Yet, in the search for an answer, the question gets lost among technological formulae and ecological cycles.

Small wonder. The rude awakening to large-scale hazards has elevated the laboratories' technical language to the rank of Kantian categories. What used only to serve in scientific experiments, but remained completely outside everyday experience, has become a precondition for everyday survival. Kant today, revised for the developed civilization of hazards, reads: 'Without becquerels one is blind to danger; with becquerels one is in thrall to a stranger.' A grid of new technological hieroglyphs has been superposed on our perception of the world, and from these in turn the cry goes up, with apparent technological inexorability, for technological solutions. Even where thought immerses itself in natural connections, so as to fight for their preservation, it surfaces with the most 'natural' idea in the world: to reorganize society in accordance with ecology's 'loose-coupling model'.

The thesis of this book, however, is that natural destruction and large-scale technological hazards can and must be apprehended and deciphered as mystified modes of social self-encounter, twisted outwards and reified. They are objectified memories of suppressed, social-human imperfection and responsibility, projected onto nature and technology. It is not something external but itself that society encounters in the hazards that convulse it; and the reigning paralysis can only be overcome in so far as society apprehends the hazards as signposts to its own history, and to its corrigibility.

The issues in human genetics are not merely those of health and medical innovations, which do not leave, but merely go further down, the road of the technically neutral and familiar. Rather, social principles, prejudices and values can now be imposed in the darkness of the womb, directly intervening in subjectivity. Under the fulsomely legitimated label of preventive health care, medical progress is turned to eugenic ends. The old dream of feasibility can avail itself step by step of the biological selection and shaping of humankind and society, in the abstract medium of laboratory practice; and it does so highly effectively, barely controllably and with the blessing of 'free will'. This diabolical world would have to be harped by angels overnight for the dams of legal loopholes so far erected or planned against this deluge of genetic engineering to delay the eugenic age into which we have already slid.

Ecological protest does not issue 'naturally' from despoliation. Dying forests and songbirds do not metamorphose, in accordance with the laws of reincarnation, into protesting humans. Despoliation and protest, rather, are isolated from or related to one another through cultural symbols, whose effectiveness today has its origin and basis in human traditions and living conditions. In other words, people who protest do so against a perceived threat, not to the environment, but to their social habitus.

The momentum of industrial development is not predicated upon a 'natural law' of civilization ('exploitation of capital', 'functional differentiation', 'bureaucratic rationality' or whatever the catchphrase may be), but also and essentially upon an extreme inequality of the burdens of proof. The flagrant contradiction in which they stand to the principle of equality is only exceeded by the unquestioning acceptance this contradiction meets with. According to the prevailing relations of definition, one side has to prove what it can never prove, in order to be given a hearing; conversely, and on the basis of the same unprovability of the consequences of their actions, the other side can do as it pleases. It is not nature but society that stabilizes the 'environment': or, more precisely, the grossly unequal distribution of uncertainty and opportunities, allowing the destruction of something *qua* 'environment' because systemic unprovability guarantees that the culprits will not be convicted. Even where they claim to be radical, ecological and technological critiques remain integrated into the prevailing, unequal distribution of burdens of proof and rules of attribution. The naturalistic fallacy shared by ecological critique and its industrial opponents is politically paralysing, concealing key dimensions and starting points for political action in risk society.

It is not the specialist logic of technology that compels us to accept hazards, but the system of organized non-liability, which renders all resistance idle, ultimately turning that which controls the production of hazards – law, science, administration, policy – into its accomplices. The rules of attribution according to which the hazards produced within the system are dealt with – i.e. calculated, justified, brought to public attention or simply rendered anonymous and palmed off on individuals (causality, guilt, liability) – stem from a different century than the hazards they are supposed to help contain socially. The result is a theatre of the absurd. In the real-life cabaret of the technocracy, the part of buffoon is played according to the rules of blind man's buff. But most amusing of all is the fact that hardly anybody notices. On the contrary, it is precisely the players who do stand no chance who fall over themselves to recognize the rules, hoping against hope to win the game thereby.

It is not the menace of annihilation, not the pollution that the media report in a kind of ongoing scandal, but distinctive features of society: in a completely administered world, superelevated safety standards and bureaucratic claims to perfection turn hazards that pass through the finest technological sieves into an internal threat to social rationality and systems. On the one hand, economy, law, science, policy with their present constitution and aims are not in a position really to dam up and forestall the hazards; on the other, the institutionalized safety pledge they furnish constitutes the embodiment, as it were, of the non-existence of hazards. Thus proof of the hazard becomes a proof of institutional failure. Reality, unreality and the political content of large-scale hazards condition one another.

The hazard itself is mute, technological, probable or improbable. It enters the world incognito, as it were, having been born of the compulsions of progress; and is swaddled in the promises of utility that stood godfather at its birth. Yet the automatism of progress finds its opponent in the hazard that has become independent, and which stands in contradiction to the prevalent fundamental norms. The mass media, new markets and social conflicts between fractions of capital in winner and loser regions intensify the spotlight that reveals hazards, in the face of their institutionalized concealment. The institutions sizzle, as it were, in the purgatory of their safety pledges, which they must themselves continually renew and fan into flame as they see them blatantly negated by accidents. This objectified counterforce of the involuntary self-revelation of hazards is admittedly predicated on social framework conditions that few countries have satisfied until now: parliamentary democracy, (relative) independence of the press and a strong economy, in which at least for the majority of the population the invisible threat of cancer is not outweighed by acute malnourishment and starvation.

It is not technology that guarantees safety, but technology in combination with social institutions and rules that render the socially produced hazards accessible to social provision and participation. The institutions – political, legal and industrial – are prisoners of their safety technology. Through the complicity of policy and technology, every suspected accident becomes a political scandal. Policy finds itself, so to speak, under house arrest in the edifice of technological safety guarantees thanks to its ceding of power to technology, while the engineers have long since left by the back door of probabilistic safety, which is by definition fallible. There is a total lack of that social understanding of safety which would guarantee attribution and remedial action; this turns a policy which acts only on the recommendation of selected experts into a lightning conductor for civil protest across party lines.

Science will not deliver us from our distress; it is not the impartial third party, incorruptibly promoting safety where others have long since fallen prey to their own interests or fantasies. There is, rather, a contradiction within science, hitherto unacknowledged and not thought through; between experimental logic and large-scale technological hazards. Science, whose judgement is the medium in which all must seek enlightenment about the reality of hazards, at once turns partisan in the dispute over the 'poisoned cake' of social wealth. Where the jeopardization of everybody has become a precondition for research, the public can no longer be kept from participating in the scientific episteme. In the age of large-scale hazards, scientific advances have perforce led science into a marriage of hatred with the public. Science's deprofessionalization of itself is the expansion opportunity for politics and the public. The hazards exacerbate the dependence of everyday life on science, but they simultaneously open the scientific monopoly on truth to public discussion, down to the details of production of the results.

Under conditions where 'nature' is industrially consumed and marketed, what appears and is dealt with as 'threats to nature' destroys not only the 'environment' but also property, capital, jobs, union power, the economic foundation of entire industries and regions. Ultimately the concrete manifestations of hazards are as irreversible and regionally identifiable as the hazards are abstract. What is denied accumulates in geographical sites, loser regions, which have to foot the bill for the destruction and its non-attributability by surrendering the very basis of their economic existence. The consequence is that the nation-state structure of political systems and ecological conflict situations of vast extent become independent of each other and give rise to geopolitical displacements. These present the economic and military bloc structure both within and between states with wholly new kinds of burdens and opportunities. We are on the threshold of a new phase of risk-society politics; in the context of disarmament and the relaxation of East–West tension, the apprehension and practice of politics can no longer be national but must be international, because the social mechanism of hazard situations flouts the nation-state and its systems of alliance. Thus apparently fixed political, military and economic constellations are set in motion, and this compels or makes possible a new global domestic policy in Europe.

This perspective, which strips large-scale hazards of their external 'environmental' character, discovering their suppressed sociality in the emergence, perception and political turbulence of these hazards, may be delimited and more precisely specified in contrast to two predominant lines of thought: 'naturalist objectivism about hazards', and 'cultural relativism about hazards'. True, hazards require natural-scientific categories and measuring instruments in order to be 'perceivable' at all, a subject of social controversies and prevention. It is equally correct to say that hazards always remain dependent on standards of tolerance which can vary from culture to culture, from group to group, indeed from one day to the next. Yet technological naturalism about hazards ignores the rule-governed character of its own definitions, and for precisely that reason chains itself to the extreme inequality of the burdens of proof.

Conversely, cultural relativism about hazards tends, by stressing the manifold cultural perceptions of hazard, to ignore the special features of large-scale technological hazards. These, unlike their pre-industrial forebears, are the result of decisions, and hence their causes and perpetrators cannot be politically neutralized by recourse to something external. They stem from human, more precisely from industrial, thinking and errors. With their advent, risk calculations break down, along with their principles of rationality (which had assumed the non-existence of large-scale hazards), even where state authority, invested billions and technical competence are shattered in the process. Cultural relativism thus ignores the social objectivity of hazard, which manifests itself in the contradiction between the safety spiral and the legalization of annihilation hazards in the technological welfare state.

Thus two global developments presently coincide, as they will continue to do, in that highly industrialized temple of progress, the German nation: on one side, a level of safety and a bureaucratic claim to control and perfection that are unprecedented, penetrating social life to its smallest crevices; and, on the other, the introduction and justification ('legitimizable' only in the transition from one century to another) of annihilation hazards which no epoch or culture has previously known, as product and reverse side of its own creative powers. This contradiction is at the centre of a social theory of hazard, of which, in conclusion, certain political consequences will be highlighted.

The politics of hazard: the principle of the indivisibility of health and life

Hiroshima was terrible – it was horror pure and simple. But there it was the enemy who struck. 'What would happen', Patrick Lagadec asks, 'if terror were unleashed from the centre of the productive, and not the military, domain of society? What industrial policy would follow the first civilian holocaust? What crisis of technology, of democracy, of reason, of society?' (1987, p. 230).

These questions indicate the centre of political force of technology's annihilation hazards: the suspicion of the extreme contrary that falls upon the guarantors of state and technological safety, and which is harder to shake off with every passing accident. This threat to all life does not come from outside; it does not begin with the exceptional case, war, the enemy. It emerges within, enduringly, as the reverse side of 'progress', peace and normality. Those responsible for safety and rationality, and those who threaten it most, are no longer isolated from one another by national or group boundaries, polarized by the roles they play, but are potentially one and the same; because of the ultimately ineradicable possibility of a catastrophe; and because of social suspicion, which is fed by catastrophic events.

This historically unprecedented, institutionally reinforced symbiosis of the defenders and jeopardizers of the highest good, this extreme two-facedness of state–technological–industrial authority – and not the hazard alone – makes clear what is at stake politically in risk civilization.

Advanced industrialism in the confusion of centuries is now toying with life, with the lives of entire regions and generations. This becomes particularly scandalous socially where the value of life, the protection of living things, is ranked high or highest in the hierarchy of values. This need not always be the case, as we know from painful experience of the Nazi past in Germany. The bare fact of the danger (though only in so far as it casts a shadow upon the custodians of order and progress), the contradiction in which it stands to the life-norm thereby injured, adverts to the origin of the hazard in decisions, and hence to its avoidability in principle, and to its comprehensive and global force. These are the components, the flammable matter of which the political dynamite of technological hazards is made.

Enduring global injuries to the basic rights of life and health cannot be legitimated in the social security milieu of Germany, and are thus not politically sustainable. One might call this the principle of indivisibility of life and safety. Whereas equality is socially divisible, and can be injured without causing instability – the most blatant present example being mass unemployment, though also past and present poverty – this does not hold for global threats. It remains unthinkable that a government Office of Life similar to the government Institute of Labour should be set up, keeping public account of the diseases and deaths fomented by industry, periodically publishing child mortality rates in hazardous regions. If millions of cancer sufferers are officially identified, then this – in contradistinction to mass unemployment – is possible only where unprovability is a foregone conclusion; otherwise it amounts to a collapse of the political system.

Material and social opportunities (education, income, property, etc.) can be distributed extremely unequally and at the same time legitimately in the 'meritocratic society'. This is guaranteed by the undoubtedly ambiguous and suspiciously ideological principle of 'just rewards', which takes as its model the examination system, transforming equality of opportunities into 'legitimate' inequality. In this sense, the educational system is the central factory of the justifications of social inequality in modern society. According to yardsticks of 'individual performance', it demonstrably and accountably transforms individual equals into unequals (in rank, salary, etc.), in such a way that the disadvantaged accept their disadvantage in the name of the principle of equality (according to the model).

There appears to be a principle of similar magical power in the domain of norms for the jeopardization of life, in the case of the 'pefectly normal' mass deaths on the streets. Traffic fatalities in Germany alone are on a scale to wipe an average village off the map every year, yet they are accepted without a collective outcry; they halt neither the boom in car sales nor the construction of expressways, while cars continue to tear along the motorways without restriction by the authorities (although everything is open to decision).

The figures resemble those of wartime casualties. Yet the apparent similarity is deceptive. The horror of mass deaths on the roads has hitherto been normalized socially for at least two reasons. First, because the risk is in principle freely taken: people can choose for themselves not to drive; and second, because a traffic accident remains an individually attributable event, thereby remaining interpretable within the categories of individual guilt. Both are excluded in the case of large-scale technological-ecological hazards. Here, despoliations and dangers to life are assigned *en bloc*, and so can neither be individually avoided nor blamed upon the victims themselves. In other words, a century-wide chasm separates 'everyday war' on the roads from modern large-scale hazards. The latter exclude the individual from decision-making, and are global in scope. Unlike traffic fatalities, large-scale hazards cannot be socially contained or excluded. Simultaneously, even where the repercussions are publicly admitted, all the instruments for individual attribution break down, along with the instruments which societies have at their disposal, and this results in the legitimation of permanent injury to the life-norm.

That means that, at their core and by virtue of their logic, conflicts about progress are conflicts about disclosure. The question of attribution is central. However, responsibility for consequences is thus covertly shuffled off. Superficially it is a matter of numbers and formulae which are supposed to uncover or conceal the wretched situation, whereas the gravity of the consequences is obvious to everyone, though usually left unspoken. The demonstration of the hazard coincides with the sounding of the alarms and the compulsion to take counter-measures. Different rules of acknowledgement and attribution might – in principle – transform a whole officialdom of denial into activists for prevention. With that, there is a division of roles and positions in the conflict over progress: on the one hand, the compulsive denials of officialdom, which are presented on the brink, as it were, of an abyss of only technically minimizable large-scale hazards; and the central significance of the unveiling of hazards and their active attribution, on the other hand, which can exploit the contradictions immanent to the legalization of hazards.

This is the explosive of which conflicts over progress are made: it is a matter of technology, of abstraction, apparently a game with numbers and formulae. Yet these are only, so to speak, the lids, which are supposed to seal the political pressure-cooker that large-scale hazards can turn a country into, and to prevent it from whistling and exploding. The immediate subject of dispute is not whether to take the steaming pot off the hob. That is certainly at issue, but is displaced into the erection and demolition of safety façades, against a background of institutionalized claims which get the bureaucratic machinery working, one way or another. What follows is an attempt to think through and extrapolate two lines of development, which do not necessarily coincide with political intentions, programmes and parties.

First, the likelier alternative: the real-life cabaret of 'more of the same', with its involuntary-intentional nightmarish perspectives of a technocratic dismantling of democracy; and second, the politics of detechnocratized enlightenment, which aims to change the rules of hazard definition and can be understood as a strategy to activate industrial self-contradictions.

To continue thus will lead to a collapse. To demand a collapse is to preserve. In other words, there will be a change of political system, one way or another. It will come either through the normality conspiracy of the side-effects, no longer unseen, of human genetic, nuclear, chemical and ecological hazards; or through an active policy of manufacturing attribution and liability in the system of organized non-liability. The opportunities, quantum leaps of development, mechanisms of both variants of future development are perhaps not even mutually exclusive; but they configure a real opposition, which determines the forthcoming confrontations about the 'politics of progress'. The following sections will give a rough sketch of the alternatives.

On the road to an authoritarian technocracy

Industrial society's answer to the risk society is the intensification of technocracy. Safety lacunae are filled – and torn open – by better and better safety technologies (including psychological and social techniques). The secular error is celebrated, initially, by oversight; then perhaps strategically, because it is liberating to exchange one compulsion for its contrary. Also, the growing political power of hazard leads to a growing call for the state to take authoritarian measures, which defend its own authority while technical safety constructs disintegrate. To the failure of the engineers is added that of policy, law, science; investments turn into losses of billions and markets collapse. This domino effect forces the state into a kind of defensive aggression, in which some day perhaps even talking about hazards will be punishable.

There are many routes to technocracy: the smooth, the noisy, the surreptitious. The last of these is realized behind the walls of non-attributability, in a reduplication of reality sealed off from everyday experiences: these are submitted to the judgement of technocracy, to its foundations and fundamental errors, which few experiences, if any, can controvert. Certainly, hazards refute technocratic rationality, but its refuted upholders hold sway over everyday life. One indication of this is the speed with which common parlance has soaked up nuclear acronyms: rem, GAU [the German initials for 'worst-case nuclear scenario']. This new language, indicating a stage as yet utterly uncomprehended in the forced scientization of the everyday, at once symbolizes a reduction of social habitus to technology that is excelled only by the vacuousness that supervenes. The technologists have not troubled themselves over the cultural and social consequences, nor has that been their task. Yet the whole world lives and thinks in a terminology connoting technological mastery and economic utility – that is, one that is predicated on modalities of action that have nothing to do with, say, the contamination of milk and vegetables. It is a language called into question, in the eyes of the world, by the catastrophes to which these modalities owe their supremacy. Thus the technologies' constructions of rationality collapse – but everyday life moves in; it has to live in them.

This can take place if, and for as long as, it remains unnoticed. The power of technocracy consists in its overruling of the senses, and thus in its overruling of the judgement of citizens in the advanced civilization of hazards. Democracy ends, not with a bang but with a hushed transition to authoritarian technocracy. The *citoyen* will perhaps not even notice that the core issues of survival have long since eluded civic participation.

In the context of all that, the confusion of centuries is not only a hindrance but can also be of service; it can be used to cushion the rationality and legitimacy of the merely technical philosophy of safety against its enemies in reality and against rebellious social movements. Nothing is so self-evident as what has, until now, always been self-evident and right. It is not inactivity but overactivity that is cultivated. Time and again, one seeks individual perpetrators. The fact that they evade discovery, let along being convicted, while the oceans and their inhabitants continue to die in spite of sweeping successes in environmental policy: that proves nothing in the end, or only the inextricability of 'fateful linkages'. In other words, the secular error guarantees both action and inaction. A great deal happens, but nothing changes. There is almost a causal relation between activity and inconsequentiality. This predestines the confusion of centuries to become a key strategy of gesture politics. It can be plausibly demonstrated to a fearful population that no danger inheres in contaminated milk powders, by having the minister responsible slurp some in front of rolling cameras with a beatific smile on his face and not dropping dead on the spot; similarly whole legions of bureaucrats and scientists can swarm out in search of 'causes' and 'culprits' (with and without success), provided that only the familiar, historically superseded rules underlie their investigations.

At first glance, a policy that replaces action with symbolic action will find that many of the conditions at hand are supportive of it. With the growth and ineluctability of the hazards, it becomes simply unbearable to know about hazards that in any case contradict one's own perceptions. Ultimately, one no longer wants to know the unpleasant truths which turn the conduct of one's own life upside down. Precisely because the dangers – the political and social dangers – are dawning on us, there is once again the threat of those who point them out being demonized into the actual danger.

Yet this alliance with the impalpability and unavoidability of the threat is extremely unstable and self-endangering. Disasters, near-disasters and suspected disasters expose to public view, and thus render fragile, the technological backwardness of policy and law. This strikes to the marrow of the security of progress in risk societies. Performing a high-wire act near catastrophes of unimaginable extent, risk societies tend to undergo basic political mood-swings. Phases of forced normality alternate with dazed states of emergency. Calls for the 'strong hand' of the state to avert ecological catastrophe tie in with calls for the 'strong hand' that is supposed to counteract the collapse of state rationality and power. At the end of it there stands an 'industry behind bars', and it is not by chance that its rituals of reinforcement and control resemble the exertions of former enemy states, promising to 'protect' their citizens from one another.

Prophetic talk of 'technology as the enemy within' (Bloch 1954, vol. 4, p. 814) is already becoming a palpable reality today, in the reinforced nuclear power plants. Only a tractable little recession – as many covertly believe – will help counteract this, so that peace and order will at last be restored to investments under the silent coercion of economic conditions.

Ecological technocracy is the daughter, or perhaps the unwanted, unloved grandchild, of ecological protest. In order to preserve what is dearest and most worthy of preservation, one yields up what is no less needful of protection, the delicate plant of democracy. Conservation can falsely play off a flayed nature's rights against human rights, and thus endanger the very thing that facilitated its breakthrough, the freedom to hold and act on contrary opinions. Today, bureaucracy is already attending to patches of reclaimed land with a solicitude that stifles, just where the grass and birds can at last breathe again. Perhaps the ecologically administered nature museums will soon need barbed-wire fences. Then, across all the entrenched oppositions and generation gaps, one may be able to organize an exchange of opinions in Germany on the subject of barbed-wire fences, as well as associated justifications and avowals of fidelity to principles.

Thus what is difficult to grasp and to bear is that the empowered technocracy can also avail itself of the in-fighting between its opponents. It has two foster-fathers fighting and unbalancing one another, and for precisely that reason the question of defensive action seems almost hopeless.

The groundwork has been set for an authoritarian technocracy. Then a light appears at the end of the tunnel. An approaching train? Or the first glimmers of a different modernity? Surely it is still permitted to ask questions and to lay traps.

Ruses of powerlessness: the utopia of a responsible modernity

The question of questions is: how can one entice, from the force of compulsive progress, the contrary power of liberation – without stasis, without the reanimation of times past, whose attractiveness derives substantially from their being past? How, then, can stone-age industrialism be shaken off, its disastrous second-order nature overcome, where it is so predominant that what are almost the only alternatives are turning a blind eye or cheering it on? How can industrial rationality be convicted of the contradiction that it represents, and has silenced in its abstract self-consciousness, so that traduced rationality will lead the rising against its betrayal in word and in deed?

How can proceedings be instituted wherein the Enlightenment is both plaintiff and defendant, with the goal of finding the way to an enlightened, tamed 'industrialism' (which would then no longer be such), conscious of its repercussions and hazards? Or, to put the question in more mechanical terms, while we are still in the age of the car, how can a steering wheel and brakes at last be installed in the driverless vehicle of the technological economy?

How can chemistry be transformed into the saviour of nature – which is what it claims, must claim, to be – or at least partly transformed by appeal to the other half of its bad conscience, also known as market instability? How can lawcourts, ministers, engineers (note the order of rank) be prevailed upon to give enough of a chance to that protection of life and safety to which they sacrifice their last drop of blood, their last fragment of thought – enough, at least, not to extend the lists of endangered species along with the lists of their successes? How, then, can the most self-evident of pledges, which has always been kept and is unquestionably central to all activity, be employed against the grievous damage perpetually inflicted on it, so that at least something will happen, and the worst be prevented?

More, new, better technology is promised. It has always been promised, and perhaps it is indispensable. It helps as much as ordering the installation of a stack of tea strainers in order to detoxify the seas. Laws pile up, not, of course, only new and better ones, but also laws that irrefutably confirm the determination to take political action. Determination to do nothing? For the problem of attribution, the burdens of proof, the possibilities of whisking up a froth of activity without anything happening are still not addressed.

This time it is not only a matter of technology or of laws, but of forestalling the storm-tide of ever new hazards as and before they arise, through a reversal of the burdens of proof. The rules allow the traffic in hazards to continue 'unseen', 'unforeseeably'. Then they send out swarms of powerless officials, virtually blind to the hazards and furnished with spades and shovels, to bring those hazards, which are unprovable and certainly unattributable, to bay: these rules of the real technological satire must be abolished, rewritten.

Only if the consequences are debated before the decisions that produce them are taken; if the injured no longer have to run an obstacle course of impossibilities, but the perpetrators are compelled to prove that their production and products are non-hazardous: only then will that which has always seemed actual become possible, a mode of production in which there is a chance of knowing what one is doing.

This revolutionary change in the relations of definition thus has a powerful ally, namely the fact that it has always been happening – according to the pledges of those to be disempowered. We ought to take them at their word. People have always denounced what is already on the verge of collapse. This is the counter-paradox to the futility of a Kafkaesque resistance, which starves while being held at arm's length by accepted unprovability.

It is not merely a new move in the game of chess, a new strategy in a familiar game, but a different set of rules of the game: the redistribution of the burdens of proof, different standards and principles for the scientific and legal judgement and condemnation of hazards. Thus interventions are necessary at the foundations of industrial production, science, law and politics, in order to render possible rationality, responsibility, decision-making, participation, democracy.

What it takes is an eye for the rule-governed character of what is apparently natural, natural-scientific, an eye for how the points have been set, making what is necessary necessary; and for ways of adjusting them, so that the train of argument and action will travel in the opposite direction. That will require flagrant outspokenness, and a good deal of rage; cunning, a great deal of that, and some laughter too. As paper is patient, I shall now set out how this is to be achieved, or at least the general direction.

One would expect scant hope of success from a 'missing person' notice in search of the revolutionary subject, whether in the columns or over a full page of the tabloid, quality or underground press. Naturally, it does one good to appeal to reason as rigorously as one can, and for that reason alone it can do no harm – because, as we know from experience, it makes little impression from a realistic point of view. A group may yet be founded for solving the world's problems – and the understanding of the political parties is devoutly to be wished for. Certainly the 'power' of social movements, 'citizens' commitment', the adherence to the letter of a 'critical mass of the public' – that is, everyone to whom the appeal has gone out for a better future – are indispensable. If, however, contrary to expectations, this should prove insufficient to achieve the leap from stone-age, catastrophist industrialism to the enlightenment and serenity of a responsible modernity, then just this once there remains an extra prescription (which actually I may only whisper). What must be done is to mobilize the strongest ally of the authoritarian technocracy, and draw it over to the side of life and the future – and that ally is the technocracy itself. The prescription is as clear as daylight. We only lack the faith that it can be realized.

Not to go too far into the details, the idea might be elucidated as follows: the concealed uprising (with which the technocracy conspires) can be contained by the counter-forces and opposing powers that it must mobilize. The technocracy must strike a blow against itself, with its left hand not knowing what its right hand is doing; this will make it do what it has always claimed to do, but which is pretty exactly the contrary of what is perpetually happening in its internal strife, which has become independent.

It is thus a question of ruses of powerlessness, whose power may consist in uncovering the symbiosis, rife with contradictions, of safety and its contrary, and in playing out the schizophrenia of supremacy against itself, and thus promoting the victory of the institutionalized claim of rationality and self-determination over its institutionalized betrayal. At its core is the strategic insight that even the most technocratic technocracy is really as imperfect as the behaviour it displays. It is only a loosely concealed collection of contradictions, which ossifies in its claim to perfection.

To put it another way, the danger that cannot be allowed to exist has internally split the technocracy (bureaucracy, science, the economy and of course politics) into two or more groupings (of enthusiasts for progress, dissidents against it, sickly opportunists and those who want to be able in future to tell their children and girlfriends where and how they earn their daily bread without feeling sick to their stomachs). But it has furthermore split the technocracy into one authority which invokes the law, ardently declaring on the occasion of every refutation by accidents that the danger does not exist (or more precisely, cannot exist because otherwise it would exist); and into the other authority which equally invokes the law, creating, expanding the danger through the declaration of its normality.

From wherever they are pursued – in the workplace, in citizens' intiatives, in editorial offices, the lawcourts, parliament, or government – policies that oppose the technocracy of hazards must be able to play like a keyboard the compulsions and contradictions of a hazard normalization which has become independent. It must score the arguments and actions of the technocracy in order to elicit, from the instruments of their denial, the melody of restraint and release from hazards and from technocratic trusteeship. The preceding chapters contain some preliminary hints on the requisite arrangement and pitching of the keyboard of arguments, so that by merely crooking one's fingers the dead wood of denial is transformed into the resonant tones of counter-action.

Every emergent hazard elicits from industry and from the administration the pontifications and conjurations of technocratic infallibility. These must be encouraged, and transformed into opportunities for the popes of technocracy to expose themselves by means of their own claims to rationality, until, instead of the sumptuous robes of their certainties, only naked, human uncertainty remains. The pompous, affected claims of safety can also be confronted with the baser reality of their 'quite normal' failures. One can expose the *absolvo*, with which the technocracy acquits itself of the sins it daily commits; it can be known, and made known, with an astonishment educated and refined by the repeated eliciting of its claims. With everybody's agreement, one can apprise the siblings of industry and politics of the horns of their dilemma: namely, either to practise the chemistry of the pure charity that they preach, or, by admitting the contrary, to sound the sirens of danger in plain view of everyone; and thus be compelled, as the implementers of pure safety, to switch off what they have covered up hitherto.

Three types of situation and practice can be distinguished – albeit unsystematically, only by way of example, and somewhat crudely.

An initial possibility is, first, to declare how tolerant the lords of the technocratic monopoly are in acquitting themselves of their normal, everyday failings. Even against the background of a technological conception of safety, agreement with the technological trusteeship on safety issues entails agreeing that wear and tear, faults – in short, the whole gamut of technocratic imperfection – can be turned against the technocracy itself. It means using them as an opportunity for what is enforced by safety flaws in highly hazardous areas: shutdowns. If this point of acceptance is switched, the train goes off in the opposite direction: strategies for the denormalization of acceptance.

This still relates to the awakening as it were of the technocratic conscience to its practices. We must distinguish two such strategies which break up technological monism and the technologists' monopoly on safety issues by imposing on them the yardstick of an extended safety definition, the very yardstick implied by the technocratic truncation of 'safety' to the preservation of life. These are strategies of de-monopolization, and of an extended safety definition.

Yet readjusting people's understanding of safety is not the only way to tease out the contradictions and political power of publicly legalized hazards. In the third strategy, the banners of liability, com-

petence and responsibility that are held aloft everywhere must be
shown to take precedence over the practices of non-liability, incompetence and irresponsibility. This can be achieved by unashamedly
practising what the prevalent power relations exclude: strategies of
redistribution of burdens of proof and of the manufacture of
attributability.

These opportunities for resistance by the powerless, to which the
sociological term 'strategies' has been applied here, share one feature:
the only weakly disguised self-contradiction of the technocracy,
which reveals itself precisely with technological perfection. The technocracy has to declare its safety under the thumbscrews of an alarmed
public, while unprecedented, purely technological and hence irreduc-
ible annihilation hazards are legalized, jeopardizing the central value
of world society, the lives of everyone.

The 'technocratic absolvo': the disclosure and revocation of the sanctioning of normal technological imperfection

The most complex technical formulae and calculations will hardly
invalidate a rule that is admittedly greatly oversimplified, unlikely to
win any scientific laurels and yet famous throughout the world:
Murphy's Law. Generally speaking, this can be boiled down to the
formula, 'Whatever can go wrong will go wrong, and even what can't
go wrong will do so.' This norm of technical hitches, errors, minor
and major accidents, etc. presents the politically interesting problem
of how, in the domain of annihilation hazards, the normality of errors
is commensurable with the constant improvement, which simultaneously becomes a necessity, of technological safety.

Precisely here there lies a substantial starting point for a counter-
politics which refuses to accept the conventional divide between the
worlds of everyday flaws and those of super-safety, a politics which
thrusts the reality of technologists' inadequacies in the field of safety
under the nose of their claims to perfection.

Such strategies for the disclosure and revocation of the sanctioning
of normal technological imperfection can be carried out particularly
effectively in the upper echelons of the political executive, management and research. They are effective and legitimate because the
rejection of safety flaws of all kinds in the high-safety realm is assured
of widespread public approval. Conversely, the unacknowledged ac-

ceptance of faults, while the sworn oaths of safety are renewed, does not exactly look like the icing on the cake of virtue.

A good illustration of this strategy is afforded by the procedure of the Kiel provincial government, which in summer 1988 refused, on account of 'safety-related flaws', to license the reopening of the Brokdorf nuclear plant after a replacement of thermal elements. The minister responsible referred his decision to an inspector's report of the Technical Surveillance Association. This had determined that one of some 360 centring rods had broken in the reactor pressure chamber. It also cited the damage caused to eighteen carbon rods during thermal element replacement, from which small projecting corners and suspension elements were now missing. The technical experts themselves had pointed out this fault, but did not recommend that the plant be shut down. The minister, on the other hand, took the view that 'according to my understanding of nuclear plant safety, something like this is completely out of the question.' (*Süddeutsche Zeitung*, 17 August 1988.) Having said this, he refused to reconnect the nuclear plant to the electric grid, and cast the opposing party's federal minister in the role of bogeyman: either accept a blatant fault at the plant – and you will publicly raise your hand to swear that it is safe – and issue a directive for the plant to be reopened, or accept and enforce the arguments that contradict your own pro-nuclear policy. The Federal Energy Ministry promptly replied that such faults were perpetually arising, though none until now had called for such far-reaching measures. Finally a directive was issued, and the plant was reopened. The consequence was that responsibility for the population's safety shifted from regional to federal government. It is not hard to imagine the political dispute that would flare up if a failure were to occur at the plant.

Questions of acceptance are simply outside technology's brief, even and especially where the technologists have hitherto been able, on the strength of their monopoly, to absolve themselves of their sins. An executive wanting to free itself from the role of puppet for the basic technological policy decisions of experts can win back political independence by exposing such internal technological contradictions.

Yet the risk of this strategy is inseparable from its advantages. It stays within the circle of technological safety, finally accepting the authority of the technocracy even to refute itself. For instance, it remains unclear whether the courts, too, can or will take the step towards their juridical autonomy.

*Safety against safety, or how the technological monopoly
can be cracked with the lever of its
own premises*

The only instrument against the technological monopoly is the one
which it has always used itself: the dramaturgy of safety. This can
only threaten the technocracy, however, if it is based on a 'philosophy' of safety which abolishes the prevalent equation of safety with
technological safety.

Safety is not safety, but a definition subject to rules and interpretations which can and must be changed if the supposed safety demonstrably produces and establishes extreme unsafety. Thus the
technological safety monopoly can be questioned from two sides:
first, by demonstrating that technological rationality is in principle
unsafe; second, by showing that safety can never be ensured by
technical means. This is all, first, in the interest of safety; and second,
in that of the technicians, who also only ever have safety on their
minds anyway.

Technological safety must exclude what can never be technically
excluded: the worst-possible-case scenario. It cannot exclude the
latter because all its calculations are built on the quicksand of probabilities and correspondingly have two ineradicable blemishes in the
age of annihilation hazards. First, even the least likely worst-case
scenario, one of which will turn up every million years, can arise
tomorrow. And second, even predictions of high statistical significance remain untested, as they must, for even a theoretically discounted worst-case scenario cannot affect the validity of probabilistic
theories, even if it extinguishes life.

If one insists on technological safety, one must rule out worst-case
scenarios, because only thus can the technological monopoly on
safety be justified. That means that technologists under the guillotine
of refutation are compelled to conceal the uncertainties of their
calculations, and to assert the safety of their technological systems
come hell or high water. Their safety monopoly drives them into a
dogmatic mania for perfection, and precisely as 'perfection' comes
under excessive strain, opportunities are multiplied for public demonstrations of the insecurity of technical certainties and security. This
takes place in the interplay of expert opinions, as exact as they are
divergent; in the quick change of 'advances in knowledge', now in one
direction, now in another; in the open plurality of risk studies; and in
the style of wrestling that will of necessity decide the outcome of the
struggle.

To the self-contradictoriness of technocracy there corresponds a gain in the power of policy. Technical decisions need not have technical grounds. There's the democratic rub. Loosening the grip of the technocratic monopoly can tease out and aid the revival of dormant but not dead spirits of irrationalism; that is one of the reverse sides of this development.

Yet anyone who argues and fights within the prespecified definitions and rules remains trapped in the circular labyrinth of mere technological definitions and controls, with whose bolt-holes and dungeons the technologists are increasingly well acquainted. They can always entice the critics into dead ends and let them waste away on a diet of bread and water. Conversely, however, strategies for an expanded definition of safety can lay bare the overextended and extremely shaky construction of mere technological safety. This change of conceptual standpoint alone brings unexpected advantages. More makes one conscious of less: the inadequacy of mere technological safety is revealed, at the centre of technological perfection.

One can take up the case of one social institution after another: economic restitution; attributability; medical, cultural or religious provision. Everywhere the result is the same: protection from catastrophe is to the threat of catastrophe as a sticking plaster is to a mushroom cloud. That, however, exposes a yawning gap in prior provision in the welfare state, annulling its rubber-stamped safety claim quite independently of catastrophic occurrences.

All this, however, changes nothing of the fact that technologists and medical specialists have the say on safety matters, and represent the government and judiciary – at least if one refers to the small print, the 'specifications for implementation'. The empowerment formula of 'the state of science and technology', to which everyone swears allegiance as the basis for decision-making, elevates the scientists to the throne of technological safety in hazard civilization. The opening-out of the definition must be made concrete through a social opening of the advisory bodies and offices of standards, an open breach of the social monopoly. It must involve the inclusion of experts and counter-experts, finely balancing a variety of disciplines, so that their systematic errors throw one another into relief. Perhaps one could even stipulate the participation of 'lay judges' according to prescribed rules, which one might crib from the institution of the jury in a court of law.

In any case, the modes of participation and organization exclude the possibility of groups of experts adjudicating over whether to acknowledge and remove their own errors. A division of power is required between research and applications, between the diagnosis of hazards and therapeutic safety measures. The 'executive' which commissions operations ought to be separated from the 'legislature' which stipulates the rules of proof and decision-making. The latter, in its turn, would be distinct from a 'scientific judiciary' for regulating complaints and individual cases, according to prescribed ground rules. Only in the light of these possibilities does it become clear that we live today in a kind of 'dictatorship of progress'.

How the perpetrators can be released from their predestined anonymity: the redistribution of burdens of proof and the production of attributability

Those who accept the inequality of the burdens of proof submit themselves to the predetermined unprovability of their surmises. If it is abolished or reversed, two things come to consciousness: the existence of the inequality, and its lack of justification. Here too, it is true that other, wider norms of hazard definition not only make one conscious of the basic faults and inequalities in prevalent norms; they simultaneously constitute the yardstick for a critique of prevalent norms. To turn the tables, forcing the perpetrators to prove, before they unleash on humanity their half-baked 'knowledge' (which will be mouldy be tomorrow), that the repercussions are inconsiderable and unhazardous, for example in human genetics, means that experiments and applications can be carried out only where progress and safety, which have always tripped so lightly off the tongue, are confirmed in accordance with prescribed rules.

This apparently minute change is far-reaching in its assumptions and consequences. The burden of insecurity and unpredictability is transferred to the hazard-producers, in this case, the molecular biologists and reproduction engineers. The latter are compelled to think ahead, beyond the boundaries circumscribed by their confederacy of dunces, and to do something completely unheard of, to avail themselves of a multidisciplinary approach. Uncertainty is reintroduced into research. This would compel in-depth surgery, cutting current practices to the quick. Research freedom, now predicated on the freedom to apply its results, would have to be reorganized. At the least, responsibility for consequences, in so far as this is possible, could be made the subject of an expanded and renewed logic of discovery.

In this process, not only the foundations of science and technology are subject to compulsive change; different foundations for industrial research applications, and different legal foundations for research and jurisdiction, are also required. Everywhere, reversal of the burdens of proof would compel a more modest, 'pro-visionary', 'pre-visionary', and thus decidable, responsible ethos amid the emergence and dissemination of hazards. Nor would this occur in regional bureaucracies, or far away in parliaments, but at the very site of the incident, in the form of preventive checks and self-controls during the genesis of hazards. The question of causal connections would be addressed to the specialists.

For industry, even a partial reversal of the burden of proof would lead to a complete change of production policy. Before they could endanger the world in the name of progress, the perpetrators would have to present everything in accordance with a strict procedure. All the evidence which the injured parties have hitherto been obliged, and almost never been able, to muster against the information monopoly and the causal and legal paradigm errors, would have to be provided by the perpetrators in accordance with certain rules – before they can endanger the world under the seal of approval of progress. The common decency of not poisoning others, a rule that ought not to require any special codification or justification, precipitates a collapse of (power) relations under conditions of standard legalized pollution. The burden of insufficient proof, the complaints, suspicions, drama-tizations with which people respond to suspected poisoning – all this would pile up in managers' offices and bring to a speedy end the recklessness that serves the interests of capital. Caution would be the mother in the kitchen of toxins. Only those who could really substantiate the paradise proclaimed by their press departments and full-page glossy advertisements could assert themselves in the marketplace.

Non-attributability, which protects the perpetrators, does not by any means protect the injured parties. On the contrary: non-attributability is a highly one-sided mode of distribution of burdens of proof. People compelled by their natural location to swallow poisons have also to suffer the declaration, framed with every admin-istrative skill, that the situation they are in does not exist. At the same time, however, this means that the extreme unfairness of non-attributability contains a lever for producing attributability.

The destruction that takes place is 'beyond provable attribution' only if the law and causality are interpreted in individual terms. A form of thought based on correlations and broad perspectives, for example, would indeed allow the quantities, kinds and sources of poisons to be determined. A regional correlational framework is always capable of making attributions. Even if it is unclear which plant, or which person in which plant, has opened the sluices to pollutants, it will hardly be claimed that the fish and the fishermen produce the poison that robs them of their livelihood and lives. There are industrial regions that profit at the expense of tourist and agricultural regions. Even the bitter experience that this logic is not recognized in current law does not avail against this insight. Thus there are manifold possibilities for producing principles of attribution and liability even where the established rules refuse to.

Here, too, different rules mean a different reality. Not only because 'reality' is the product of rule-governed definitions, but because the very possibility of different rules of attribution throws light, as appalling as it is true, on the ones that prevail. The prevailing rules perfect the powerlessness of the counter-force, and discharge their poisonous freight, extremely one-sidedly, into other industries and regions.

Ecological democracy

While the technological imagination is perpetually spurred on and valorized, its social and sociological counterparts find themselves curiously restricted to the here and now. Technologies have been given the go-ahead to turn the world upside down. Even such changes in the constitution of life, for example, as slip through with human genetics, are by no means compelled to legitimize themselves. It is only at some later time that one wonders that the family has become extinct, like dinosaurs and ladybirds. Yet anyone daring to reflect on whether this inexhaustible cornucopia of technological world improvements also requires a social and social-scientific opponent and fellow player, soon falls under the shadow of suspected treason. With a few exceptions, admittedly, sociologists, political scientists and legal specialists have continued to exercise self-censorship. Silently, and with the blessing of an abstinence enjoined by professional ethics, they turn themselves into the mouthpieces of a 'now' whose having-become is almost as indiscernible as the becoming that acts within it.

In political theory, a programme to deal with large-scale hazards might require revolutionary changes. Until now, these have been sought and discussed principally in two ways: through a change in the majority rule (Claus Offe) or through the creation of a constitution that is capable of learning (Ulrich Preuss).[1]

Four phases of rationalization and development are generally distinguished in the development of the modern bourgeois constitutional state under the rule of law. The first is the neutralization of parties in conflict and the stabilization of social peace, seen as a response to the seventeenth-century civil wars of religion. State authority is achieved by establishing peace at home. The second phase is characterized by the guarantee of liberty, in which individuals are released from the compulsions and habits of feudal society. Third, in modern political and constitutional theory a universalization of the injunction to equality is established and drawn up, and is expressed above all in universal and equal suffrage. Fourth and last, the principle of fraternity, of solidarity, becomes one of the most recent principles of state organization, in the form of the welfare state with its encompassing human provision (Habermas 1986–9, vol. 2; Guggenberger 1984).

These four forms of development are now called into question, because they do not touch upon the foundations of human dominion over nature, and cannot of themselves banish 'the threatening shadows of civilizatory inhumanity, indeed of the relapse of the entire species into barbarism' (Guggenberger and Offe 1984, p. 15, quoted in Demirovic 1987, p. 97).

Doubts about majority rule are based, first, on its lack of legitimacy among the minorities it works against, fomenting rather than resolving social conflicts. Second, it is criticized for reducing the complex material on whose grounds ecological matters are decided to the all but irrelevant alternatives of yes or no. Accordingly, some suggested reforms try to articulate majority rule in a realistic, i.e. applicable, way, and open the system to the consideration of minority interests. Perhaps the most far-reaching suggestion is that of Claus Offe, who would apply the majority rule to itself.

> It is by no means agreed that the majority would always decide in favour of a majority decision. It is much more plausible and imaginable, even with the greatest reservations, that there would be a greater or lesser deviation from this schematic rule, according to the distribution of costs and benefits involved in a specific issue that is to be decided – say, in establishing moratoriums, catalogues of eliminated possibilities, quotas, the taking of a second vote, issue-specific redefinitions of constituency boundaries, the delegating of decisions to specific bodies, the redefinition of voting rights by categories of person, and so forth. (1984, p. 181)

For Ulrich K. Preuss, the task of an ecology-orientated constitutional and democratic theory consists in making time, and in keeping the future open to democratic options through the power of a minority veto. Above all, however, the constitution must be enabled to learn. Like Offe, Preuss considers an open procedure and future guaranteed only if a reflexive organization of the constitution is accomplished.

> A constitutional theory would then no longer be only a theory concerning the institutional framework for determining social contradictions and antitheses, but also a theory concerning conditions for the process wherein society perpetually revolutionizes its own basic structures. The constitution itself would then be continuously changing, so that one might say that a constitutional theory would be a theory of controlled constitutional change. (1985, p. 78)

Both suggestions, however, seem at once 'too elevated' and 'too immanent' from the perspective presented here. In other words, they change too much and too little at once. Indeed, a change in the majority rule creates a new situation in general, not only with regard to ecological conflicts and their solutions. At the same time, it remains uncertain whether rules for minorities automatically and sufficiently guarantee the inclusion of criteria for damage to one's interests.

Finding and inventing a responsive constitution constitutes the counterpart to the reflexivity of industrial modernity. In other words, it is indispensable. Yet the mere capacity to learn remains empty; it does not prescribe what is to be changed or when, or what the priorities are. Furthermore, both suggestions are situated within the political system, remaining within the framework of previous definitions of politics and non-politics, thus failing to recognize the central fact that hazard situations arise from the connection between economy and science, economy and law, economy and state. The recommendations thus posit as constant what must be revealed and posited as variable, so that injured (minority) interests can effectively be brought to bear on them: the prevailing relations of definition, and the relations of power which are inscribed in them (and in which they are inscribed). These strike the nerve between politics, the economy, law and science; they redistribute participatory rights, opportunities for control; indeed they make it possible to transform the anarchy of progress into actual decisions, which cannot be reached in the prevailing conditions.

Political development in hazard civilization is approaching the crucial issue of the redistribution and democratic shaping of the principles, rules and foundations of the power to define terms: different relations of proof, different relations of restraint, different relations of control and guidance, different relations of participation in decision-making. The intermeshing of subsystems under predetermined, inexorable progress can be disengaged by a network of reciprocal control measures. By holding the debates about repercussions before their conversion into reality, by enhancing the possibilities of contradiction, the tempo of development is reduced, thus institutionally making possible attribution and the capacity to learn.

Not postmodernity but enlightenment against industrial society

The character of industrial society is such that its momentum contradicts self-determination, as fatalism contradicts democracy, and organized non-liability contradicts rationality and justice.

Undoubtedly the whole European tradition can be swept away at a stroke and thrown into the moth-eaten trunk of the nineteenth century. That is postmodernity, the reign of cynicism – which can no longer be named and recognized as such. The 'brave new world' can supervene, if and because the cultural horizon, on which it still appears and can be criticized *qua* 'brave new world', has been fragmented and expunged.

Alternatively one can recognize how stubborn, marvellously unresponsive and half-baked modernity is. It has spurred on and made concrete the project of the Enlightenment, but also damaged and betrayed it; it has laid it out on the Procrustean bed of a minimalism rife with contradictions, of inconsequential participatory decision-making, rampant technocracy, universal threats to life and systemic dangers. At least it has done so in thought; in reality, self-sufficient modernization has long since begun to break up the supposedly compulsory union of industrial society and modernity. Thereby it also offers the opportunity to take the Enlightenment out of mothballs, as a social movement and political force against industrial fictions and narrow-mindedness. It is not the end of the Enlightenment, but its deployment against industrial society here and now, that is on today's agenda, although few people pay it much attention.

A final question

What if radioactivity gave one an itch? Realists, also known as cynics, will answer that something would be invented, for example an ointment to 'turn off' the itch. It would be big business. Certainly there would soon be explanations, which would be highly effective among the public, that the irritation meant nothing, that perhaps it was caused by phenomena other than radioactivity. Harmless in any event; unpleasant, but demonstrably not malignant. One would assume that such excuses would have scant chance of survival, with everyone walking around covered in rashes and scratching inflamed skin, from models on photographic assignments to board members at the United Institute for Denial. The excuses would be scratched out, as it were. Nuclear policy, as well as dealing with large-scale modern hazards, would be confronted with a completely changed situation: the subject under dispute, the subject at hand, would fall within the orbit of cultural experience. The consequences of progress would not only injure people, but this injury would also be an unpleasant experience engraved on their minds.

This is what will decide the future of democracy: are we dependent on the experts for every detail in issues concerning survival, or does the culturally manufactured perceptibility of hazards restore to us the competence to judge for ourselves? Are the only alternatives now an authoritarian or a critical technocracy? Or is there a way of counteracting the disempowerment and expropriation of everyday life in hazard civilization?

Notes

Chapter 1   Barbarism Modernized: The Eugenic Age

1   Benda Committee (1985) on in-vitro fertilization, genome analysis and gene therapy.
2   The 1958 Nobel prizewinner, J. Lederberg, famously defined the human genotype as a molecular sequence, 180 cm in length, of carbon, hydrogen, oxygen and phosphorus atoms – that is, DNA – formed in a dense spiral of 5 billion nucleotide pairs, in the nucleus of the original ovum and that of every mature cell (Lederberg 1988, p. 292). Even Testart, human-gene engineer and a critic of his profession, says, 'It only gave me a weird feeling at the beginning . . . it doesn't any more. *The eggs that begin to live and separate always look identical*' (1988, p. 65; my italics).

Chapter 2   The Naturalistic Misunderstanding of the Green Movement: Environmental Critique as Social Critique

1   This relates to a classical problem of subject–orientated sociology, taking for the field of reference of its analysis the horizon of the individual agents (cf. Bolte 1983).
2   'Man is a child gone astray in the "forests of symbols" (Baudelaire),' says Kundera (1988, p. 63). Cf. Douglas and Wildavsky (1982) on the cultural dependence of risk perception generally. I am grateful to R. M. Lepsius who, in discussion at the University of Heidelberg, insisted on the importance of acceptance.
3   I show in chapter 4 that revocation of acceptance also reflects institutional contradictions in the security state.

4   'In the early 1960s the majority of citizens still believed that problems of
    environmental pollution, such as air and water purity, could be solved
    through technological progress. Today only a minority believe that, and most
    people associate technology generally with environmental damage' (Klipstein
    and Strumpel 1984, p. 66). A questionnaire commissioned by *Der Spiegel* in
    February 1988 showed that that a stable 79 per cent of the population
    opposed nuclear power, and wanted to see the last nuclear plant shut down
    within ten years at the latest.

5   This argument is an attempt to clarify the connection between parts I and
    II of my *Risk Society* (1992), i.e. the 'jeopardy thesis' and the 'individualiz-
    ation theorem'. The connection was barely made there, as many critics justly
    noted.

6   'In his economic culture today the German citizen feels exhausted and
    satiated, not proud but exiled; increasingly shut in by the growing mountains
    of consumer goods, which are no longer attractive and only conceal the
    prospect of a better life – a greyer atmosphere than in any other of the western
    industrialized countries investigated', wrote Michael von Klipstein and
    Burckhardt Strümpel in the conclusion of a comparative international study
    (1984, p. 128).

Chapter 3   Industrial Fatalism: Organized Irresponsibility

1   See Luhmann (1989, esp. pp. 165ff) for a challenging reformulation of this
    proposition.

2   Compare the detailed account in my *Risk Society* (1992), chapter 7.

3   As the representative Shell study (Fischer, Fuchs and Zinnecker 1985), *Jugend
    und Erwachsene '85* ('Youth and adults '85') shows, 46 per cent of the young
    see a dark future, while 54 per cent are more optimistic. Yet this picture is
    deceptive, as the authors of this investigation demonstrate. For the optimists
    have a tendency to conform, are comparatively unconcerned about politics
    and tend to take flight in superficial leisure activities. In contrast, the pessi-
    mists are surprisingly more active in politics, take an interest in new forms of
    energy and environmental protection and are also more frequently ready to
    engage in these issues in their spare time.

Chapter 4   The Self-refutation of Bureaucracy:
The Victory of Industrialism over Itself

1   In a representative survey carried out in 1987 of the West German popu-
    lation's expectations for the future, the following question was asked: 'I shall
    now name several areas of life. Please indicate the areas in which you would
    like to become more active in future.' Nine per cent of respondents said
    'becoming more involved in political and union matters', while 55 per cent
    replied 'living more healthily'.

2 Much too little attention has been paid, in my opinion, to the numerous arguments that both the bureaucratic phenomenon and Weber's model of bureaucracy are in some respects specifically German. After all, the smooth running of a bureaucracy of the hierarchical type is predicated upon an authoritarian culture. Hannah Arendt gave a particularly penetrating account of this collusion or opposition between a country's political culture and its bureaucratic servility in her book *Eichmann in Jerusalem* (1979), with reference to the persecution and extermination of the Jews. Thus, for example, the Danes under military occupation gave a lesson in civil disobedience that was never properly learned in Germany. The yellow star constituted a line of social demarcation, the preliminary stage before the transportation and extermination of the Jews; this was subverted by the royal family's declaration to the baffled army of occupation that they, like every other citizen, would wear theirs with pride. The degree of autonomization of the bureaucracy, its 'machine-like character', is also the objectified, institutionalized obverse of a submissive culture that has faith in authority; and for which thinking and acting outside state regulations is at best suspect, and normally a threat to the state, but always incomprehensible.

3 This is how the comprehensive empirical study by Fischoff et al. summarizes the issue of the acceptability of risks: 'Not only do we not know how safe is safe enough, but our analysis also led us to the opinion that this question is vaguely formulated or misconceived. There are no universally acceptable risks as such, but only risks accepted as the result of a determinate approach and mode of procedure, reflecting a specific group's way of seeing related matters and statements of the problem at the time of the decision' (1981, p. 24). In risk research, the rule of thumb is that we voluntarily expose ourselves to hazards around a hundred to a thousand times greater than those imposed on us from outside. Most scientists consider an annual individual death risk of one in a million, say, from the effects of industrial plants or chemicals, to be tolerable. This constitutes a markedly lower value than that posited for known natural risks, such as diseases or natural disasters. It is further assumed that an 'appropriate' utility, for its part admittedly subject to personal and sociocultural variations, is associated with the risk.

4 Dieter Schäfer did the groundwork on the social regulation of questions of liability for ecological hazards. Kaufmann's basic idea is that the problem of systemic security cannot be solved by technological means alone.

Chapter 5 Implementation as Abolition of Technocracy: The Logic of Relativistic Science

1 The growth of science has taken on extreme dimensions. 'Eighty per cent of all scientific and technological discoveries hitherto, and over 90 per cent of the total scientific and technological information in the world, was produced in the twentieth century; of this, over two-thirds has emerged since the

Second World War. The generation of scientists alive today comprises some 80 per cent of all scientists who have ever lived' (Kreibich 1986, p. 126). The latest research findings show particularly high growth figures, according to which scientific and technological information increases by 13 per cent annually, thus doubling every $5\frac{1}{2}$ years – and the trend is upwards, because of the development of increasingly powerful information systems (ibid., p. 27).

2   'The threat is not only to jurisprudence as the dominant paradigm for interpreting norms for over 300 years, but also to the classical political elite as the agent for setting standards. What is ultimately at stake is the "primacy of law" as a politically posited norm, in face of the norms established by committees, to which it is unconstitutional to grant supreme political authority. Finally, it also affects the position of the lawcourts as the final court of appeal, because the roles of judge and expert can no longer be performed according to the envisaged juridical procedure. The experts are thus liberated from their "advisory capacity" to the degree in which the judges feel out of their depth on complex matters of technical safety' (Wolf 1987).

3   This is not only metaphorically true: the lawcourt, which has been condemned to reach a decision, is ground between the wheels of formal responsibility and material incompetence in disputes over technological safety; meanwhile the technicians, who actually hold sway, meekly assume the role of specialist advisers. This game of switched roles is negotiated according to the model of 'engineering judgement': judges project judge-like qualities – relevant knowledge, impartiality, objectivity – onto the experts who *de facto* hold sway, thus passing off the latter's judgements as their own. 'Thus the specialist arrives at a self-image composed of the clichés of the judge's role. "Engineering judgement" means that engineers produce risk analyses according to research maxims of the "plausibility" of safety analyses; they "ascertain the facts", define the "worst-case scenario", adjudge the necessity of safety precautions in accordance with the principle of "balanced assessment". The engineers in turn ground this in "sound engineering sense", laying claim to "plausibility" and "unrestricted freedom to act", and demanding the brief of determining the "acceptance of socially acceptable risks. Assuming related patterns of belief between the judge and the specialist, it would then seem advisable, in order to preserve the judge's role, "not to delve too deeply into technical-scientific problems in the evaluation of complex matters" and "not to deviate from the statements of the specialists who are capable of judging technological borderline cases". That, however, means that the natural and technological sciences have set up extremely successful sub-offices in the areas traditionally taken care of by ethics, law, and politics. And it is precisely this thesis that indicates the line of resistance formed by all those who understand ethics, law and politics to be a principle of opposition to technological development' (Wolf 1987).

Chapter 6   The 'Poisoned Cake': Capital and Labour in Risk Society

1   See especially the works of K.-H. Ladeur, K. M. Meyer-Abich, K. U. Preuss, A. Rossnagel, E.-H. Ritter, G. Winter, T. Blanke and K. Bosselmann, besides the books by R. Wolf.

Chapter 7   Conflicts over Progress: The Technocratic Challenge to Democracy

1   The debate is still in the preliminary stage, with the thesis yet to be published.

Bibliography

Adorno, T. W. (ed.) 1969: *Spätkapitalismus oder 'Industriegesellschaft'?* Frankfurt

Altmann, R. 1987: *Der wilde Frieden.* Stuttgart

Alvares, C. and Rannesh, B. 1987: 'Damning the Nevada: The Politics behind the Destruction'. *The Ecologist* 17, 62–73

Améry C. 1978: *Natur als Politik: Die ökologische Chance des Menschen.* Reinbek

Améry, J. 1980: *Jenseits von Schuld und Sühne: Bewältigungsversuche eines Überwältigten.* Stuttgart

Anders, G. 1980: *Die Antiquiertheit des Menschen: Über die Zerstörung des Lebens im Zeitalter der dritten industriellen Revolution*, 2 vols. Munich

Anders, G. 1983: *Die atomare Bedrohung.* Munich

Andrews, W. 1987: 'Reaktionen der Verwaltung auf grossflächige Gefahrenlagen: Krisenmanagement'. In C. Böhret et al., 59–64

Apitz, K. 1987: *Konflikte, Krisen, Katastrophen: Präventionsmassnahmen gegen Imageverlust.* Frankfurt

Arendt, H. 1970: *Macht und Gewalt.* Munich

Arendt, H. 1979: *Eichmann in Jerusalem*, revised edn. Penguin, Harmondsworth

Augstein, R. 1977: 'Atomstaat oder Rechtsstaat?' *Der Spiegel* 10, 29

Bacon, F. 1980: *New Atlantis.* In *The Great Instauration*, ed. J. Weinberger. Arlington Heights

Bätzing, W. 1984: *Die Alpen: Naturbearbeitung und Umweltzerstörung: eine ökologisch-geographische Untersuchung.* Frankfurt

Barker, E. 1985: 'New Religious Movements: Yet Another Great Awakening'. In P. E. Hammond (ed.), *The Sacred in a Secular Age*, Berkeley, 36–57

Bechmann, A. 1987a: *Öko-Bilanz.* Munich

Bechmann, A. 1987b: 'Die ökologische Herausforderung'. In J. Hesse and C. Zöpel (eds), *Zukunft und staatliche Verantwortung*, Baden-Baden

Bechmann, G. (ed.) 1984: *Gesellschaftliche Bedingungen und Folgen der Technologiepolitik*. Frankfurt

Bechmann, G. 1987: 'Sozialwissenschaftliche Forschung und Technikfolgenabschätzung'. In K. Lompe (ed.), *Techniktheorie, Technikforschung, Technikgestaltung*, Opladen, 28–58

Beck, U. 1983: 'Jenseits von Stand und Klasse?' In R. Kreckel (ed.), *Soziale Ungleichheiten, Soziale Welt*, Special issue 2, Göttingen

Beck, U. 1985: 'Von der Vergänglichkeit der Industriegesellschaft'. In T. Schmid, *Das pfeifende Schwein*, Berlin, 85–114

Beck, U. 1992: *Risk Society: Towards a New Modernity*, translated by Mark Ritter. London

Beck, U., Giddens, T. and Lash, S. 1994: *Reflexive Modernization* – Politics, Tradition and Aesthetics in the Modern Social Order, Cambridge: Polity Press

Beck, U. and Beck-Gernsheim, E. 1995: The Normal Chaos of Love, Cambridge: Polity Press

Becker, E. and Ruppert, W. (eds) 1987: *Ökologische Pädagogik: Pädagogische Ökologie*. Frankfurt

Becker, U. (ed.) 1982: *Staatliche Gefahrenabwehr in der Industriegesellschaft* (Schriften der Deutschen Sektion des Internationalen Instituts für Verwaltungwissenschaften 6). Bonn

Beck-Gernsheim, E. 1987a: 'Fortschritt ohne Masstab? Sozialwissenschaftliche Argumentationen zu den neuen Fortpflanzungstechnologien'. *Sozialer Fortschritt* 10, 235–8

Beck-Gernsheim, E. 1987b: 'Ganz normale Familien? Neue Familienstrukturen und neue Interessenkonflikte durch Fortpflanzungstechnologien'. In B. Lutz (ed.), *Technik und sozialer Wandel: Verhandlungen des 23. Deutschen Soziologentages in Hamburg 1986*, Frankfurt, 277–92

Beck-Gernsheim, E. 1988a: *Die Kinderfrage: Frauen zwischen Kinderwunsch und Unabhängigkeit*. Munich

Beck-Gernsheim, E. 1988b: 'Zukunft der Lebensformen'. In J.-J. Hesse, H.-G. Rolff and C. Zöpel (eds), *Zukunftswissen und Bildungsperspektiven*, Baden-Baden, 99–118; English translation: *Technology, the Market and Morality* (in press)

Bell, D. 1973: *Die Zukunft der westlichen Welt: Kultur und Technologie im Widerstreit*. Frankfurt

Benjamin, W. 1978: *Illuminations*. New York

Benn, G. 1982: *Gedichte* (text of the first edn). Frankfurt

Berger, J. 1986a: 'Gibt es ein nachmodernes Gesellschaftsstadium? Marxismus und Modernisierungstheorie im Widerstreit'. In *idem* (ed.), *Die Moderne: Kontinuitäten und Zäsuren, Soziale Welt*, Special issue 4, Göttingen

Berger, J. (ed.) 1986b: 'Die Moderne: Kontinuitäten und Zäsuren', *Soziale Welt*, Special issue 4, Göttingen

Berger, J. 1988: 'Modernitätsbegriffe und Modernitätskritik in der Soziologie'. *Soziale Welt* 2, 224–36

Berger, P. L., Berger, B. and Kellner, H. 1975: *Das Unbehagen in der Modernität.* Frankfurt and New York

Bergmann, J., Brandt, G., Körber, K., Mohl, E. and Offe, C. 1969: 'Herrschaft, Klassenverhältnis und Schlichtung'. In T. W. Adorno (ed.), *Spätkapitalismus oder Industriegesellschaft,* Stuttgart, 67–87

Beyerlin, U. 1985: 'Die Beteiligung ausländischer Grenznachbarn an umweltrechtlichen Verwaltungsverfahren und Möglichkeiten zu ihrer vertraglichen Regelung auf "euroregionaler" Ebene'. *Natur und Recht 5,* 173–9

Biedenkopf, K. H. 1985: *Die neue Sicht der Dinge: Plädoyer für eine freiheitliche Wirtschafts- und Sozialordnung,* Munich

Bloch, E. 1959: *Prinzip Hoffnung,* vol. 4. Frankfurt

Birkhofer, A. and Lindackers, K.-H. 1980: 'Technik und Risiko'. In R. Lukes and A. Birkhofer (eds), *Rechtliche Ordnung der Technik als Aufgabe der Industriegesellschaft,* Cologne, 97–114

Biswas, M. R. and Biswas, A. K. 1982: 'Environment and Sustained Development in the Third World: A Review of the Past Decade'. *Third World Quarterly 4,* 479–91

Blanke, T. 1985: 'Recht, System und Moral: Vorüberlegungen zu einer ökologischen Verfassungstheorie'. In H.-E. Boettcher (ed.), *Recht, Justiz, Kritik: Festschrift für Richard Schmid,* Baden-Baden, 395–418

Blanke, T. 1986: 'Autonomie und Demokratie'. *Kritische Justiz 4,* 406–22

Bleck, J. and Schmitz-Feuerhake, I. 1979: *Die Wirkung ionisierender Strahlen auf den Menschen.* Bremen

Böhret, C. 1987: 'Innovative Bewältigung neuartiger Aufgaben'. In Böhret et al., 29–58

Böhret C. and Franz, P. 1982: *Technologiefolgenabschätzung.* Frankfurt

Böhret, C., Klages, H., Reinermann, H. and Siedentopf, H. (eds) 1987: *Herausforderungen an die Innovationskraft der Verwaltung.* Opladen

Bohrer, K. H. 1987: 'Nach der Natur: Ansicht einer Moderne jenseits der Utopie'. *Merkur 8,* 631–45

Bolte, K. M. 1983: 'Subjektorientierte Soziologie'. In *idem* and E. Treutner (eds), *Subjektorientierte Arbeits- und Berufssoziologie.* Frankfurt, 12–37

Bonss, W. 1988: 'Zwischen Emanzipation und Entverantwortlichung: Zum Umgang mit den Risiken der Gentechnologie'. In *Herstellung der Natur? Fragen an den Bericht der Enquetekommission 'Chancen und Risiken der Gentechnologie',* Munich

Bonss, W. and Hartmann, H. 1985: 'Konstruierte Gesellschaft, rationale Deutung: Zum Wirklichkeitscharakter soziologischer Diskurse'. In *idem, Entzauberte Wissenschaft: Zur Relativität und Geltung soziologischer Forschung, Soziale Welt,* Special issue 3, Göttingen

Bookchin, M. 1974: *Umwelt und Gesellschaft: Diskussion um Bookchin.* Hamburg

Bookchin, M. 1982: *The Ecology of Freedom: The Emergence and Dissolution of Hierarchy.* Palo Alto

Bosselmann, K. 1985: 'Wendezeit im Umweltrecht'. *Kritische Justiz 4,* 345–61

Bosselmann, K. 1986: 'Eigene Rechte für die Natur?' *Kritische Justiz* 1, 1–22

Bosselmann, K. 1987: 'Die Natur im Umweltrecht'. *Natur und Recht* 1, 1–6

Bräutigam, H. H. and Mettler, L. 1985: *Die programmierte Vererbung: Möglichkeiten und Gefahren der Gentechnologie.* Hamburg

Brand, K.-W., Büsser, D. and Rucht, D. 1984: *Aufbruch in eine andere Gesellschaft: Neue soziale Bewegungen in der Bundesrepublik.* Frankfurt

Brecht, B. 1964: *Schriften zum Theater.* Frankfurt

Brecht, B. 1967: *Gesammelte Werke*, vol. 14. Frankfurt

Breuer, S. 1986: 'Ist Umweltzerstörung überhaupt vermeidbar? Niklas Luhmann über "Ökologische Kommunikation"'. *Merkur* 8, 681–4

Brodsky, J. 1987: 'Less than One' and 'In a Room and a Half'. In *idem, Less than One*, London

Brooks, H. 1984: 'The Resolution of Technically Intensive Public Policy Disputes'. *Science, Technology, Human Values* 9, 1

Brückner, P. 1972: *Freiheit, Gleichheit, Sicherheit.* Frankfurt

Bühl, W. L. 1981: *Ökologische Knappheit: Gesellschaftliche und technologische Bedingungen ihrer Bewältigung.* Göttingen

Bühl, W. L. 1986: 'Soziologie und Systemökologie'. *Soziale Welt* 4, 363–89

Bühl, W. L. 1987: 'Grenzen der Autopoiesis'. *Kölner Zeitschrift für Soziologie und Sozialpsychologie* 2, 221–54

Bundesminister der Justiz (ed.) 1987: *Der Umgang mit dem Leben: Fortpflanzungsmedizin und Recht.* Cologne

Bundesminister für Ernährung, Landwirtschaft und Forsten 1986: *Waldschäden in der Bundesrepublik Deutschland: Ergebnisse der Waldschadenserhebung 1986.* Münster-Hiltrup

Bundesminister für Umwelt, Naturschutz und Reaktorsicherheit (ed.) 1984: *Umweltradioaktivität und Strahlenbelastung: Jahresbericht.* Bonn

Burckhardt, J. 1984: *Briefe.* Basle

Burke, E. 1953: *Reflections on the Revolution in France: And on the Proceedings in certain Societies in London relative to that Event.* London

Burnheim, J. 1967: *Über Demokratie: Alternativen zum Parlamentarismus.* Berlin

Caldwell, I. K. 1985: *International Environmental Policy: Emergence and Dimensions.* Durham

Campbell, D. T. 1985: 'Häuptlinge und Rituale: Das Sozialsystem der Wissenschaft als Stammesorganisation'. In Bonss and Hartmann, 257–74

Cansier, D. 1975: *Ökonomische Grundprobleme der Umweltpolitik.* Berlin

Carson, R. L. 1962: *Silent Spring.* New York

Chargaff, E. 1986: 'Der Kunstgestopfte Schleier der Maja: Betrachtungen zur Gentechnologie'. *Merkur* 8, 664–75

Clausen, L. and Dombrowski, W. R. (eds) 1983: *Einführung in die Soziologie der Katastrophen.* Bonn

Comfort, L. (ed.) 1987: *Disaster Management.* Durham

Commission of the European Communities 1986: *The Europeans and their Environment in 1986.* Brussels

Commoner, B. 1963: *Science and Survival.* New York

Conrad, J. 1978: *Zum Stand der Risikoforschung*. Frankfurt

Conrad, J. 1987: 'Risikoforschung und Ritual. Fragen nach den Kriterien der Akzeptabilität technischer Risiken'. In B. Lutz (ed.), *Technik und sozialer Wandel: Verhandlungen des 23. Deutschen Soziologentages in Hamburg 1986*, Frankfurt, 455–63

Crenson, M. A. 1971: *The Un-Politics of Air Pollution: A Study of Non-Decisionmaking in the Cities*. Baltimore

Crozier, M. 1964: *Le Phénomène bureaucratique*. Paris

Czada, R. and Drexler, A. 1988: 'Konturen einer politischen Risikoverwaltung: Politik und Administration nach "Tschernobyl"'. *Österreichische Zeitschrift für Politikwissenschaft* 1, 52–66

van den Daele, W. 1985: *Mensch nach Mass? Ethische Probleme der Genmanipulation und Gentherapie*. Munich

van den Daele, W. 1986: 'Technische Dynamik und gesellschaftliche Moral: Zur soziologischen Bedeutung der Gentechnologie'. *Soziale Welt* 2/3, 149–72

van den Daele, W. 1987: 'Politische Steuerung, faule Kompromisse, Zivilisationskritik: Zu den Funktionen der Enquetekommission "Gentechnologie" des deutschen Bundestages'. *Forum Wissenschaft* 1, 40–5

Dahrendorf, R. 1980: *After Social Democracy* (Liberal Publication Department, Unservile State Papers 25). London

Darnstädt, T. 1987: *Gefahrenabwehr und Gefahrenvorsorge*. Frankfurt

Demirovic, A. 1987: 'Staat und Technik: Zum programmatischen Charakter politischer Theorie bei Carl Schmitt und Ernst Forsthoff'. In H. Loewy and T. Kreuder (eds), *Konservatismus in der Strukturkrise*, Frankfurt

Der Rat von Sachverständigen für Umweltfragen 1987: *Kurzfassung des Umweltgutachtens 1987*. Bonn

Deutsche Forschungsgemeinschaft 1987: 'Stellungnahme zum Bericht der Enquete-Kommission "Chancen und Risiken der Gentechnologie" des 10. Deutschen Bundestages' ('Statement on the report by the commission of inquiry into "Risks and Opportunities of Genetic Engineering'. Tenth German Bundestag (manuscript)

Devereux, G. 1967: *Angst und Methode in den Verhaltenswissenschaften*. Munich

Dierkes, M., Edwards, S. and Coppock, R. (eds) 1980: *Technological Risk: Its Perception and Handling in the European Community*. Königstein

Dietz-Will, A. 1982: 'Fallstudie zum Konflikt zwischen Ökonomie und Umwelt'. *WSI Mitteilungen* 12, 776–8

Dörre, K. 1987: *Risikokapitalismus: Zur Kritik von Ulrich Becks 'Weg in die andere Moderne'*. Marbach

Dost, B. 1983: *Die Erben des Übels*. Munich

Douglas, M. and Wildavsky, A. 1982: *Risk and Culture*. Berkeley

Drewermann, E. 1986: *Der tödliche Fortschritt*. Regensburg

Dubiel, H. 1985: *Was ist Neokonservatismus?* Frankfurt

Dubiel, H. 1987: 'Die Ökologie der gesellschaftlichen Moral'. *Merkur* 12, 1039–48

Duerr, H.-P. (ed.) 1981: *Der Wissentschaftler und das Irrationale*, 2 vols. Frankfurt

Durkheim, E. 1984: *The Division of Labour in Society*, translated by W. D. Hollis, with an introduction by Lewis Coser. Basingstoke

Ebermann, T. and Trampert, R. 1984: *Die Zukunft der Grünen*. Hamburg

Ebersbach, H. 1985: 'Ausgleichspflicht des Staates bei neuartigen immissionsbedingten Waldschäden'. *Natur und Recht* 5, 165–70

Eisenbart, C. and Picht, G. 1978: *Wachstum oder Sicherheit?* Munich

Elias, N. 1982: *The Civilising Process*, translated by Edmund Jephcott, 2: *State Formation and Civilisation*. Oxford

Elliott, E. D., Ackermann, B. A. and Millan, J. C. 1985: 'Toward a theory of Statutory Evolution: The Federalization of Environmental Law. *Journal of Law, Economics and Organization* 1, 2, 313–40

Emerson, R. W. 1979: 'Nature'. In *idem, Nature*, 2nd edn enlarged. Carbondale

Enzensberger, H. M. 1973: 'Zur Kritik der politischen Ökologie'. In Kursbuch 33: *Ökologie und Politik oder Die Zukunft der Industrialisierung*, Berlin, 1–42

Eppler, E. 1981: *Wege aus der Gefahr*. Reinbek

Eppler, E. 1987: 'Vom Umgang mit der Technik'. *Der Spiegel* 22, 50–1

Eser, A. 1983: 'Ökologisches Recht'. In H. Markl (ed.), *Natur und Geschichte*, Munich

Esser, H. 1987: 'Besprechung von U. Beck, Risikogesellschaft'. *Kölner Zeitschrift für Soziologie und Sozialpsychologie* 4, 806–11

Evers, A. and Nowotny, H. 1987: *Über den Umgang mit Unsicherheit*. Frankfurt

Ewald, F. 1986: *L'État providence*. Paris

Ewers, H.-J. 1986: *Zur monetären Bewertung von Umweltschäden*. Berlin

Favret-Saada, J. 1979: *Die Wörter, der Zauber, der Tod: Der Hexenglaube im Hainland von Westfrankreich*. Frankfurt

Feist, U. and Krieger, H. 1987: 'Alte und neue Scheidelinien des politischen Verhaltens: Eine Analyse zur Bundestagswahl 1987'. *Aus Politik und Zeitgeschichte*, Supplement to *Das Parlament* 12, 33–47

Feist, U. and Liepelt, K. 1987: 'Modernisierung zu Lasten der Grossen'. *Journal für Sozialforschung* 3/4, 227–95

Fetscher, I. 1976: *Überlebensbedingungen der Menschheit: Zur Dialektik des Fortschritts*. Constance

Fetscher, I. 1982: 'Ökologie und Demokratie: Ein Problem der "politischen Kultur"'. In K. M. Meyer-Abich (ed.), *Physik, Philosophie und Politik: Festschrift für C. F. von Weizsäcker*. Munich, 89–105

Feyerabend, P. 1978: *Against Method: Outline of an Anarchistic Theory of Knowledge*. London

Fietkau, H.-J. 1984: *Bedingungen ökologischen Handelns*. Weinheim and Basle

Fietkau, H.-J. and Kessel, H. (eds) 1981: *Umweltlernen: Veränderungsmöglichkeiten des Umweltbewusstseins: Modelle, Erfahrungen*. Königstein

Fietkau, H.-J., Matschuk, H., Moser, H. and Schulz, W. 1986: *Waldstreben*. Berlin (manuscript; IIUG rep 86–6)

Fischer, A., Fuchs, W. and Zinnecker, J. 1985: *Jugendliche und Erwachsene '85*. Leverkusen

Fischoff, B., Lichtenstein, S., Slovic, P., Denby, S. and Keeney, R. 1981: *Acceptable Risk*. Cambridge, Mass.

Flechtheim, O. K. 1987: 'Die Sieben Herausforderungen an die Weltgesellschaft'. *Universitas* 12, 1234–57

Foucault, M. 1977: *Discipline and Punish: The Birth of the Prison*, translated by Alan Sheridan. London

Foucault, M. 1970: *The Order of Things*. London

Friedrichs, G., Bechmann, G. and Gloede, F. 1983: *Grosstechnologien in der gesellschaftlichen Kontroverse*. Karlsruhe

Fritsch, B. (ed.) 1981: *Die Herausforderungen der 80er Jahre*. Diessenhoffen

Fuchs, B. (ed.) 1985: 'Konfessionelle Milieus und Religiosität'. In Fischer, Fuchs and Zinnecker

Gehlen, A. 1961: *Anthropologische Forschung: Zur Selbstbegrenzung und Selbstentdeckung des Menschen*. Reinbek

Gehlen, A. 1963: 'Über kulturelle Kristallisation'. In *idem*, *Studien zur Anthropologie und Soziologie*, Neuwied, 311–28

Gehlen, A. 1980: *Man in the Age of Technology*, translated by Patricia Lipscomb. New York

Gillwald, K. 1983: *Umweltqualität als sozialer Faktor: Zur Sozialpsychologie der natürlichen Umwelt*. Frankfurt

Glotta, R. 1982: *Einführung in die Industriearchäologie*. Darmstadt

Glotz, P. 1984: *Die Arbeit der Zuspitzung*. Berlin

Gluchowski, P. 1987: 'Lebensstile und Wandel der Wählerschaft in der Bundesrepublik Deutschland'. *Aus Politik und Zeitgeschichte*, Supplement to *Das Parlament* 12, 18–32

Gorbachev, M. 1986: Speeches and Writings, 1. Oxford

Gorz, A. 1977: *Ökologie und Politik: Beiträge zur Wachstumskrise*. Reinbek

Gottweis, H. 1988: 'Politik in der Risikogesellschaft'. *Österreichische Zeitschrift für Politikwissenschaft* 1, 3–15

Greifelt, W. 1986: 'Tschernobyl aus der Sicht des Katastrophenschutzes'. *Unsere Sicherheit* 32, 27–9

Griffin, S. 1987: *Frau und Natur*. Frankfurt

Gross, P. 1986: 'Die neue Macht des Schicksals'. In R. G. Heinze (ed.), *Neue Subsidiarität: Leitidee für eine zukünftige Sozialpolitik?* Opladen, 64–91

Gruhl, H. 1975: *Ein Planet wird geplündert: Die Schreckensbilanz unserer Politik*. Frankfurt

Guggenberger, B. 1984: 'Die neue Macht der Minderheit'. In B. Guggenberger and C. Offe (eds), *An den Grenzen der Mehrheitsdemokratie: Politik und Soziologie der Mehrheitsregel*, Opladen, 207–23

Habermas, J. 1986–9: *Theory of Communicative Action*, 2 vols. Cambridge

Habermas, J. 1987: *Eine Art Schadensabwicklung*. Frankfurt

Habermas, J. 1988: 'Die neue Intimität zwischen Politik und Kultur. Thesen zur Aufklärung in Deutschland'. *Merkur* 2, 150–5

Habermas, J. 1990: *The Philosophical Discourse of Modernity*. Cambridge

Hack, L. 1988: *Vor Vollendung der Tatsachen: Die Rolle von Wissenschaft und Technologie in der dritten Phase der industriellen Revolution.* Frankfurt

Häfele, W. 1974: 'Hypotheticality and the New Challenges: The Pathfinder Role of Nuclear Energy'. *Minerva* 12, 1, 303–22

Halden, S. G. (ed.) 1984: *Risk Analysis, Institutions and Public Policy.* New York

Halfmann, J. 1986: 'Autopoiesis und Naturbeherrschung: Die Auswirkungen des technischen Umgangs mit lebenden Systemen auf den gesellschaftlichen Naturbezug'. In H.-J. Unverferth (ed.), *System und Selbstproduktion.* Frankfurt

Hamman, W. and Kluge, T. (eds) 1985: *In Zukunft: Berichte über den Wandel des Fortschritts.* Reinbek

Harich, W. 1975: *Kommunismus ohne Wachstum?* Reinbek

Hartmann, H. and Hartmann, D. 1982: 'Vom Elend der Experten: Zwischen Akademisierung und De-Professionalisierung'. *KZfSS* 1, 193ff

Hauff, V. and Müller, M. (eds) 1985: *Umweltpolitik am Scheidewege.* Munich

Heine, H. 1971: *Beiträge zur deutschen Ideologie.* Frankfurt

Heine, H. and Mautz, R. 1988: 'Haben Industriefacharbeiter besondere Probleme mit dem Umweltthema?' *Soziale Welt* 2, 123–43

Held, M. and Molt, W. (eds) 1986: *Technik von gestern für die Ziele von morgen?* Opladen

Held, N. and Koch, D. 1984: *Risiko und Sicherheit: Eine Bewertungsdimension der Sozialverträglichkeitsanalyse.* Mühlheim

Henk, H.-H. 1979: *Die zivilisatorisch bedingte Strahlenbelastung.* Cologne

Hermann, A. and Schumacher, R. (eds) 1987: *Das Ende des Atomzeitalters?* Munich

Hermann, A. G. 1983: *Radioaktive Abfälle: Probleme und Verantwortung.* Berlin

Hickel, R. 1987: 'Wirtschaften ohne Naturzerstörung'. *Aus Politik und Zeitgeschichte*, Supplement to *Das Parlament* 29, 43–54

Hillesum, E. 1983: *Das denkende Herz der Baracke: Tagebücher.* Heidelberg

Hirsch, J. and Roth, R. 1986: *Das neue Gesicht des Kapitalismus: Vom Fordismus zum Post-Fordismus.* Hamburg

Hölzle, P. 1987: 'Laborkrieg um AIDS', *Evangelische Kommentare* 4, 203–5

Hörning, K. H. 1987: 'Technik und Alltag: Plädoyer für eine Kulturperspektive in der Techniksoziologie'. In B. Lutz (ed.), *Technik und sozialer Wandel: Verhandlungen des 23. Deutschen Soziologentages in Hamburg 1986*, Frankfurt, 310–14

Hohlfeld, R. 1988: 'Biologie als Ingenieurkunst: Zur Dialektik von Naturbeherrschung und synthetischer Biologie'. *Ästhetik und Kommunikation* 69, 61–74

Hohlfeld, R. and Kollek, R. 1988: 'Menschenversuche? Zur Kontroverse um die Forschung mit Reagenzglasembryonen'. In R. Osnowski (ed.), *Menschenversuche: Wahnsinn und Wirklichkeit*, Cologne, 146–71

Hohmann, H. (ed.) 1987: *Freiheitssicherung durch Datenschutz.* Frankfurt

Hollis, M. and Lukes, S. (eds) 1982: *Rationality and Relativism.* Oxford

Hondrich, K. O. 1987: 'Ein unsichtbarer Gast sitzt am Tisch'. *Der Spiegel* 21, 237–42

Honolka, H. 1987: *Schwarz-rot-grün: Die Bundesrepublik auf der Suche nach ihrer Identität.* Munich

Horkheimer, M. and Adorno, T. W. 1979: *Dialectic of Enlightenment*, translated by John Cumming. London

Hornstein, W. 1988: 'Sozialwissenschaftliche Gegenwartsdiagnose und Pädagogik'. *Zeitschrift für Pädagogik* 3, 381–97

Horx, M. 1987: *Die Wilden Achtziger: Eine Zeitgeist-Reise durch die Bundesrepublik.* Munich

Huber, J. 1982: *Die verlorene Unschuld der Ökologie: Neue Technologien und superindustrielle Entwicklung.* Frankfurt

Illich, I. 1975: *Selbstbegrenzung: Eine politische Kritik der Technik.* Reinbek

Illich, I. 1978: *Fortschrittsmythen.* Reinbek

Ismayr, W. 1987: 'Die Grünen im Bundestag: Parlamentarisierung und Basisanbindung'. *ZParl* 3, 299–321

Israel, J. 1985: *Der Begriff der Entfremdung: Zur Verteidigung des Menschen in der bürokratischen Gesellschaft.* Reinbek

Jänicke, M. 1979: *Wie das Industriesystem von seinen Missständen profitiert.* Cologne

Jänicke, M. 1990: *State Failure: The Impotence of Politics in Industrial Society.* Cambridge

Jänicke, M., Simonis, U. E. and Weigmann, G. (eds) 1985: *Wissen für die Umwelt.* Berlin

Jahn, D. 1988: 'Tschernobyl und die schwedische Energiepolitik'. *Österreichische Zeitschrift für Politikwissenschaft* 1, 43–52

Japp, K. P. 1984: 'Selbsterzeugung oder Fremdverschulden: Thesen zum Rationalismus in den Theorien sozialer Bewegungen'. *Soziale Welt* 3, 313–29

Jaspers, K. 1961: *Die Atombombe und die Zukunft des Menschen.* Munich

Joas, H. 1988: 'Das Risiko der Gegenwartsdiagnose'. *Soziologische Revue* 1, 1–6

Joerges, B. 1987: 'Technik im Alltag oder: Die Rationalisierung geht weiter'. In B. Lutz (ed.), *Technik und sozialer Wandel: Verhandlungen des 23. Deutschen Soziologentages in Hamburg 1986*, Frankfurt, 305–9

Johnson, W. and Covello, V. (eds) 1986: *The Social Construction of Risk.* Dordrecht

Jonas, H. 1984: *The Imperative of Responsibility: In Search of an Ethics for the Technological Age*, translated by Hans Jonas with David Herr. Chicago

Jonas, H. 1985: *Technik, Medizin und Ethik: Zur Praxis des Prinzips Verantwortung.* Frankfurt

Jonas, H. 1987: The Frankfurt speech in acceptance of the German Book Trade Peace Prize. *Süddeutsche Zeitung,* 12 October

Jungk, R. 1979: *The Nuclear State*, translated by Eric Mosbacher. London

Kallscheuer, O. (ed.) 1986: *Die Grünen: Letzte Wahl? Vorgaben in Sachen Zukunftsbewältigung.* Berlin

Kaufmann, F. X. 1973: *Sicherheit als soziologisches und sozialpolitisches Problem.* Stuttgart

Keck, O. 1984: *Der schnelle Brüter: Eine Fallstudie über Entscheidungsprozesse in der Grosstechnologie.* Frankfurt

Keck, O. 1987: 'The Information Dilemma'. *Journal of Conflict Resolution* 31, 1, 139–63

Kern, H. and Schumann, M. 1984: *Ende der Arbeitsteilung?* Munich

Kerner, I. 1987: 'Giftmüllsiedlung'. *Psychologie heute* (December), 37ff

Kerner, J., Maissen, T. and Radek, D. 1987: *Der Rhein: Die Vergiftung geht weiter.* Reinbek

Kessel, H. and Tischler, W. 1984: *Umweltbewusstsein: Ökologische Wertvorstellungen in westlichen Industrienationen.* Berlin

Keupp, H. 1987: *Psychosoziale Praxis im gesellschaftlichen Umbruch.* Bonn

Kevenhörster, P. 1984: *Politik im elektronischen Zeitalter: Politische Wirkungen der Informationstechnik.* Baden-Baden

Kitschelt, H. 1984: *Der ökologische Diskurs: Eine Analyse von Gesellschaftskonzeptionen in der Energiedebatte.* Frankfurt

von Klipstein, M. and Strümpel, B. 1984: *Der Überdruss am Überfluss.* Munich

Knorr-Cetina, K. D. 1984: *Die Fabrikation von Erkenntnis.* Frankfurt

Knorr-Cetina, K. D. and Mulkay, M. (eds) 1983: *Science Observed: Perspectives on the Social Study of Science.* London

Koch, C. 1987: 'Zur Moral der Genealogie'. *Niemandsland* 3, 70–82

Koch, E. R. 1985: *Die Lage der Nation 1985/86: Umweltatlas der Bundesrepublik.* Hamburg

Koestler, A. 1986: *Als Zeuge der Zeit*

Kol Peng, K. 1987: 'Die globale Umweltkrise aus der Sicht der Entwicklungsländer'. *Epd-Entwicklungspolitik* 1987 10/11, 6–8

Kollek, R. 1988: ' "Ver-rückte" Gene: Die inhärenten Risiken der Gentechnologie und die Defizite der Risikodebatte'. In H. Hohlfeld (ed.), *Biologie und die Wirklichkeit.* Munich

Koselleck, R. (ed.) 1977: *Studien über den Beginn der modernen Welt.* Stuttgart

Kramer, D. 1986: 'Die Kultur des Überlebens'. *Österreichische Zeitschrift für Volkskunde*, Neue Serie 11, 3, 209–26

Kranz, H. 1985: 'Eugenische Utopien der Menschenzüchtung: Zur Verwissenschaftlichung des generativen Verhaltens'. *SOWI* 14, 4, 278–87

Kraushaar, W. (ed.) 1983: *Was sollen die Grünen im Parlament?* Frankfurt

Kreibich, R. 1986: *Die Wissenschaftsgesellschaft.* Frankfurt

Kreusch, J. and Hirsch, H. (eds) 1984: *Sicherheitsprobleme der Endlagerung radioaktiver Abfälle in Salz.* Hanover

Krohn, W. and Weingart, P. 1986: 'Tschernobyl: Das grosse anzunehmende Experiment'. In *Kursbuch 85: GAU: Die Havarie der Expertenkultur*, Berlin, 1–25

Küppers, G., Lundgreen, P. and Weingart, P. 1978: *Umweltforschung: Die gesteuerte Wissenschaft?* Frankfurt

Kuhn, T. 1970: *The Structure of Scientific Revolutions.* Chicago

Kundera, M. 1988. *The Art of the Novel.* London

Ladeur, K.-H. 1986: 'Entschädigung für Waldsterben? Die Grenzen des Haftungsrechts und die verfassungsrechtliche Institutsgarantie des Eigentums'. *DÖV* 11, 445–54

Ladeur, K.-H. 1987a: 'Rechtliche Steuerung der Freisetzung von gentechnologisch manipulierten Organismen'. *Natur und Recht* 2, 60–7

Ladeur, K.-H. 1987b: *Risiko und Recht: Von der Rezeption der Erfahrung zum Prozess der Modellierung.* Bremen (manuscript)

Ladeur, K.-H. 1987c: 'Schadensersatzansprüche des Bundes für die durch den Sandoz-Unfall entstandenen "ökologischen Schaden"?' *NJW* 21, 1236–41

Lagadec, P. 1987: *Das grosse Risiko: Technische Katastrophen und gesellschaftliche Verantwortung.* Nördlingen

Lakatos, I. 1978/80: *The Methodology of Scientific Research Programmes,* ed. John Worrall and Gregory Currie. Cambridge

Lalonde, B. 1978: 'Kurze Abhandlung über die Ökologie'. In C. Leggewie and R. de Miller (eds), *Der Walfisch-Ökologie Bewegung in Frankreich.* Berlin

Langeder, E. and Schmidl, K. 1982: *Wissenschaft und Technik im Umweltschutzrecht.* Linz

Langer, S. 1986: 'Der Mensch im Umweltrecht'. *Natur und Recht* 7, 270–6

Lau, C. 1985: 'Zum Doppelcharakter der neuen sozialen Bewegungen'. *Merkur* 12, 1115–20

Lauber, V. 1988: 'Zur Politischen Theorie der Naturzerstörung'. *Österreichische Zeitschrift für Politikwissenschaft* 1, 79–90

Lederberg, J. 1988: 'Die biologische Zukunft des Menschen'. In R. Jungk (ed.), *Das Umstrittene Experiment: Der Mensch.* Munich

Leipert, C. 1984a: 'Bruttosozialprodukt, defensive Ausgaben und Nettowohlfahrtsmessung: Zur Ermittlung eines von Wachstumskosten bereinigten Konsumindikators'. *Zeitschrift für Umweltpolitik* 3, 229–55

Leipert, C. 1984b: 'Ökologische und soziale Folgekosten der Produktion'. *Aus Politik und Zeitgeschichte,* Supplement to *Das Parlament* 19, 33–47

Leipert, C. 1986: 'Social Costs of Economic Growth'. *Journal of Economic Issues* 20, 1, 109–31

Leipert, C. and Simonis, U. E. 1985: *Arbeit und Umwelt.* Berlin (manuscript; IIUG pre 85-7)

Lem, S. 1990: *The Star Diaries.* London (reprint)

Lenk, H. and Ropohl, G. (eds) 1987: *Technik und Ethik.* Stuttgart

Lepenies, W. 1989: *Gefährliche Wahlverwandschaften.* Stuttgart

Lompe, K. (ed.) 1987: *Techniktheorie, Technikforschung, Technikgestaltung.* Opladen

Löw, R. 1984: 'Die Aktualität von Nietzsches Wissenschaftskritik'. *Merkur* 426 (June), 399–409

Löw, R. 1985: *Leben aus dem Labor: Gentechnologie und Verantwortung: Biologie und Moral.* Munich

Lowe, P. D. and Morrison, D. 1984: 'Bad News or Good News: Environmental Politics and the Mass Media'. *Sociological Review* 32, 75–90

Lübbe, H. 1987: *Politischer Moralismus: Der Triumph der Gesinnung über die Urteilskraft.* Berlin

Luhmann, N. 1986: 'Die Welt als Wille ohne Vorstellung: Sicherheit und Risiko aus der Sicht der Sozialwissenschaften. *Die politische Meinung* 229, 21

Luhmann, N. 1987: *Die Moral des Risikos und das Risiko der Moral*. Bielefeld (unpublished manuscript)

Luhmann, N. 1989: *Ecological Communication*. Cambridge

Lukes, R. (ed.) 1980: *Gefahren und Gefahrenbeurteilungen im Recht: Rechtliche und technische Aspekte von Risikobeurteilungen, insbesondere bei neuen Technologien*. Cologne

Lukes, R. and Birkhofer, A. 1980: *Rechtliche Ordnung der Technik als Aufgabe der Industriegesellschaft*. Cologne

Lurker, M. 1967: *Der Baum im Glauben und Kunst*. Baden-Baden

Lutz, B. 1984: *Der kurze Traum immerwährender Prosperität*. Frankfurt

Lutz, B. 1987: 'Das Ende des Technikdeterminismus und die Folgen: Soziologische Technikforschung vor neuen Aufgaben und neuen Problemen'. In B. Lutz (ed.), *Technik und sozialer Wandel: Verhandlungen des 23. Deutschen Soziologentages in Hamburg 1986*, Frankfurt, 34–52

Mackensen, R. 1988: 'Die Postmoderne als negative Utopie'. *Soziologische Revue* 1, 6–12

Marcuse, H. 1986: *One-dimensional Man: Studies in the Ideology of Advanced Industrial Society*. London

Maren-Griesbach, M. 1982: *Philosophie der Grünen*, in the monograph series Geschichte und Staat: Kritisches Forum 267. Munich and Vienna

Marquard, O. 1987: 'Futuristischer Antimodernismus: Bemerkungen zur Geschichtsphilosophie der Natur'. In O. Schwemmer (ed.), *Über Natur*, Frankfurt, 91–104

Marx, K. 1975: *Early Writings*. Harmondsworth

Marx, K. and Engels, F. 1966: *Politische Ökonomie*. Frankfurt

Maus, I. 1986: *Rechtstheorie und Politische Theorie im Industriekapitalismus*. Munich

Mayer-Tasch, P.-C. 1980: *Ökologie und Grundgesetz: Irrwege, Auswege*. Frankfurt

Mayer-Tasch, P.-C. 1985: 'Die internationale Umweltpolitik als Herausforderung für die Nationalstaatlichkeit'. *Aus Politik und Zeitgeschichte*, Supplement to *Das Parlament* 20, 3–13

Mayer-Tasch, P. C. (ed.) 1986: *Die Luft hat keine Grenzen: Internationale Umweltpolitik: Fakten und Trends*. Frankfurt

McKean, M. 1981: *Environmental Protest and Citizen Politics in Japan*. Berkeley

Meinberg, V. 1986: 'Strafrechtlicher Umweltschutz in der Bundesrepublik Deutschland'. *Natur und Recht* 2, 52–60

Melucci, A. 1985: 'The Symbolic Challenge of Contemporary Movements'. *Social Research*, 798–816

Merkel, K. (ed.) 1986: *Umweltschutz in beiden Teilen Deutschlands*. Berlin

Mettler-Meibom, B. 1987: *Die sozialen Kosten der Informationsgesellschaft: Überlegungen zu einer Kommunikationsökologie*. Frankfurt

Meyer-Abich, K. M. 1986a: 'Technische und soziale Sicherheit: Lehren aus den Risiken der Atomenergienutzung'. *Aus Politik und Zeitgeschehen*, Supplement to *Das Parlament* 32, 19ff

Meyer-Abich, K. M. 1986b: *Wege zum Frieden mit der Natur: Praktische Naturphilosophie für die Umweltpolitik.* Munich 1986b

Meyer-Abich, K. M. and Schefold, B. 1986: *Die Grenzen der Atomwirtschaft.* Munich

Michal, W. 1988: *Die SPD: Staatstreu und jugendfrei*

Michelmann, H. W. and Mettler, L. 1987: 'Die in-vitro Fertilisation als Substitutionstherapie'. In S. Wehowsky (ed.), *Lebensbeginn und Menschenwürde,* Munich

Moscovici, S. 1982: *Versuch über die menschliche Geschichte der Natur.* Frankfurt

Mosdorf, S. 1987: 'Die SPD muss die Partei des neuen Fortschritts werden'. *Die Neue Gesellschaft* 2 (Frankfurter Hefte)

von den Mühlen, B. 1987: ' "Tschernobyl": Bürgerinformation? Informationswirrwarr?' In C. Böhret et al., 249–52

Müller, E. 1986: *Innenwelt der Umweltpolitik: (Ohn)macht durch Organisation.* Opladen

Müller, H. 1987: 'Umwelt und gewaltsamer Konflikt: Umweltschäden und innerstaatliche Gewalt in der Dritten Welt'. *Epd-Entwicklungspolitik* 5, 79–81

Müller-Brandeck-Bocquet, G. 1988: 'Technologiefolgenabschätzung: Ein gangbarer Weg aus der "Risikogesellschaft"? Das Beispiel kerntechnischer Folgenabschätzung in der Bundesrepublik Deutschland'. *Soziale Welt* 2, 144–63

Müller-Hill, B. 1984: *Tödliche Wissenschaft: Die Aussonderung von Juden, Zigeunern und Geisteskranken 1933–1945.* Reinbek

Müller-Rommel, F. 1985: 'The Greens in Western Europe: Similar but Different'. *International Political Science Review* 6, 483–99

Mumford, L. 1986: *The Future of Technics and Civilization.* London

Murswiek, D. 1985: *Die staatliche Verantwortung für die Risiken der Technik: Verfassungsrechtliche Grundlagen und Immissionsschutzrechtliche Ausformung.* Berlin

Naisbitt, J. 1984: *Megatrends.* Bayreuth

Nash, J. R. 1976: *Darkest Hours: A Narrative Encyclopedia of Worldwide Disasters from Ancient Times to the Present.* Chicago

Nelkin, D. and Pollak, M. 1981: *The Atom Besieged.* Cambridge, Mass. and London

Nichols, E. 1987: 'U.S. Nuclear Power and the Success of the American Antinuclear Movement'. *Berkeley Journal of Sociology,* 167–92

von Nieding, G. and Wagner, H. M. 1982: 'Prinzipien der Grenzwertfestlegung (maximale Immissionskonzentrationen) für inhalative Noxen, dargestellt am Beispiel des Schwefeldioxyd (SO_2)'. *Atemwegs-Lungenkrankheiten* 4, 190–3

Nietzsche, F. 1966. *Werke,* 3 vols. Munich

Nowotny, H. 1979: *Kernenergie: Gefahr oder Notwendigkeit.* Frankfurt

OECD 1985: *The State of the Environment 1985.* Paris

OECD 1987: *Environmental Data-Compilation 1987.* Paris

Oechsle, M. 1988: *Der ökologische Naturalismus: Zum Verhältnis von Natur und Gesellschaft im ökologischen Diskurs.* Frankfurt

Offe, C. 1972: *Strukturprobleme des kapitalistischen Staates*. Frankfurt
Offe, C. 1980: 'Konkurrenzpartei und kollektive politische Identität'. In R. Roth (ed.), *Parlamentarisches Ritual und politische Alternative*, Frankfurt, 26–42
Offe, C. 1984: 'Politische Legitimation durch Mehrheitsentscheidung?' In B. Guggenberger and C. Offe (eds), *An den Grenzen der Mehrheitsdemokratie: Politik und Soziologie der Mehrheitsregel*, Opladen, 150–83
O'Riordan, T. 1983: 'The Cognitive and Political Dimensions of Risk Analysis'. *Journal of Environmental Psychology* 3, 345–54
Oswalt, W. 1983: 'Die politische Logik der Sonnenblume'. In Kraushaar
Overington, M. A. 1985: 'Einfach der Vernunft folgen: Neuere Entwicklungstendenzen in der Metatheorie'. In Bonss and Hartmann
Pampe, J. 1984: 'Waldsterben: Ein einkommenspolitisches Problem'. *Agrarische Rundschau* 4, 14–18
Parsons, T. 1986: *Societies*. London
Pelzer, N. 1987: 'Aktuelle Probleme des Atomhaftungsrechtes nach Tschernobyl'. *Energiewirtschaftliche Tagesfragen* 37, 81–5
Perrow, C. 1984: *Normal Accidents: Living with High Risk Technologies*. New York
Perrow, C. 1986: 'Lernen wir etwas aus den jüngsten Katastrophen?' *Soziale Welt* 4, 390–401
Pestalozzi, H. A. 1985: *Die sanfte Verblödung: Gegen falsche New Age-Heilslehren und ihre Überbringer: Ein Pamphlet*. Düsseldorf
Popper, K. 1934: *The Logic of Scientific Discovery*. London
Popper, K. 1979: *Objective Knowledge: An Evolutionary Approach*. Oxford
Preuss, U. K. 1984a: *Politische Verantwortung und Bürgerloyalität*. Frankfurt
Preuss, U. K. 1984b: 'Die Zukunft: Müllhalde der Gegenwart?' In B. Guggenberger and C. Offe (eds), *An den Grenzen der Mehrheitsdemokratie: Politik und Soziologie der Mehrheitsregel*, Opladen, 224–39
Preuss, U. K. 1985: 'Aktuelle Probleme einer Verfassungstheorie'. *Prokla* 61, 65–79
Prigogine, J. 1986: 'Science, Civilisation and Democracy'. *Futures*, 496ff
Prittwitz, V. 1984: *Umweltaussenpolitik: Grenzüberschreitende Luftverschmutzung in Europa*. Frankfurt
Radkau, J. 1986a: 'Angstabwehr: Auch eine Geschichte der Atomtechnik'. In *Kursbuch 81: GAU: Die Havarie der Expertenkultur*, Berlin, 27–53
Radkau, J. 1986b: 'Tschernobyl in Deutschland?' *Der Spiegel* 20, 35ff
Rajagopal, A. 1987: 'And the Poor get Gassed: Multinational-aided Development and the State: The Case of Bhopal'. *Berkeley Journal of Sociology*, 129–52
Rammert, W. 1983: *Soziale Dynamik der technischen Entwicklung*. Opladen
Rammert, W. 1987: 'Der nicht zu vernachlässigende Anteil des Alltagslebens selbst an seiner Technisierung'. In B. Lutz (ed.): *Technik und sozialer Wandel: Verhandlungen des 23. Deutschen Soziologentages in Hamburg 1986*, Frankfurt 320–8
Reichelt, G. and Kollert, R. 1985: *Waldschäden durch Radioaktivität*. Karlsruhe
Renn, O. 1984: *Risikowahrnehmung in der Kernenergie*. Frankfurt

Richardson, J. J. and Watts, N. S. J. 1985: *National Policy Styles and the Environment*. Berlin (manuscript; IIUG dp 85–16)

Richter, H. E. 1979: *Der Gotteskomplex: Die Geburt und Krise des Glaubens an die Allmacht des Menschen*. Reinbek

Rifkin, J. 1985: *Declaration of a Heretic*. Boston

Rifkin, J. 1987: 'Welche Zukunft wollen wir?' *Natur 9*, 52–4

Ritsert, J. 1987: 'Braucht die Soziologie noch den Begriff der Klasse?' *Leviathan* 1, 4–38

Ritter, E.-H. 1987: 'Umweltpolitik und Rechtsentwicklung'. *Neue Zeitschrift für Verwaltungsrecht* 11, 929–38

Rolff, H. G. 1988: 'Zukunftsverantwortung des Staates und Zukunftserwartungen der Bevölkerung'. In J. J. Hesse, H. G. Rolff and C. Zöpel (eds): *Zukunftswissen und Bildungsperspektiven*, Baden-Baden, 34–52

Ronge, V. 1978: 'Staats- und Politikkonzepte in der sozio-ökologischen Diskussion'. In M. Jänicke (ed.), *Umweltpolitik*, Opladen, 213–48

Roqueplo, P. 1986: 'Der Saure Regen: Ein "Unfall in Zeitlupe"'. *Soziale Welt* 4, 402–26

Rosenblad, S. 1986: *Der Osten ist grün? Ökoreportagen aus der DDR, Sowjetunion, Tschechoslowakei, Polen, Ungarn*. Hamburg

Rosenmayr, L. 1988: *Älterwerden als Erlebnis: Herausforderung und Erfüllung*. Vienna

Rossnagel, A. 1983: *Bedroht die Kernenergie unsere Freiheit?* Munich

Rossnagel, A. 1984a: *Radioaktiver Zerfall der Grundrechte?* Munich

Rossnagel, A. 1984b: *Recht und Technik im Spannungsfeld der Kernenergiekontroverse*. Opladen

Roters, W. 1987: 'Innovative Reaktionen auf technologische und ökologische Herausforderungen'. In Böhret et al., 109–22

Roth, C. 1987: 'Hundert Jahre Eugenik'. In *idem* (ed.), *Genzeit*, Zurich

Roth, R. 1987: 'Auf dem Wege in die Risikogesellschaft'. *Sozialwissenschaftliche Literatur-Rundschau* 15, 19–25

Rothman, B. 1986: *The Tentative Pregnancy*. Harmondsworth

Rothschild, J. 1989: 'Engineering Birth: Toward the Perfectability of Man?' In S. L. Goldman (ed.), *Science, Technology and Social Progress*, Bethlehem, Penn.

Rousseau, J.-J. 1979: *Émile, or an Education*, translated by Allan Bloom. New York

Rucht, D. 1987: 'Zum Verhältnis von sozialen Bewegungen und politischen Parteien'. *Journal für Sozialforschung* 3/4, 297–313

Ryll, A. and Schäfer, D. 1986: 'Bausteine für eine monetäre Umweltberichterstattung'. *Zeitschrift für Umweltpolitik und Umweltrecht* 2, 105–35

Sahner, H. 1984: 'Wer fordert die parlamentarische Mehrheitsdemokratie heraus?' *Zeitschrift für Parlamentsfragen* 4, 571–6

Sanchez, V. and Castellejos, M. 1985: 'Luftverschmutzung und Luftreinhaltepolitik in Mexiko-Stadt'. *Aus Politik und Zeitgeschichte* 33/4, 25–31

Sand, P. H. 1985: 'Internationales Umweltrecht im Umweltprogramm der Vereinten Nationen'. *Natur und Recht* 5, 170–3

Sandbach, F. 1980: *Environmental Ideology and Policy.* Oxford

Schabert, T. 1969: *Natur und Revolution: Untersuchungen zum politischen Denken im Frankreich des 18. Jahrhunderts.* Munich

Schacht, K. 1987: 'Alte Partei und neue Schichten'. *Die neue Gesellschaft* 4, Frankfurter Hefte, 358–62

Schäfer, D. 1972: *Soziale Schäden, soziale Kosten und soziale Sicherung.* Berlin

Scheer, J. 1987: 'Grenzen der Wissenschaftlichkeit bei der Grenzwertfestlegung. Kritik der Low-Dose-Forschung'. In B. Lutz (ed.), *Technik und sozialer Wandel: Verhandlungen des 23. Deutschen Soziologentages in Hamburg 1986,* Frankfurt, 447–54

Schell, J. 1984: *Die Abschaffung: Wege aus der atomaren Bedrohung.* Munich

Schelsky, H. 1965: 'Der Mensch in der wissenschaftlichen Zivilisation'. In *idem, Auf der Suche nach Wirklichkeit,* Düsseldorf and Cologne, 439–80

Schindler, N. 1985/6: 'Jenseits des Zwangs?' *Zeitschrift für Volkskunde* 81/2, 192–219

Schmid, T. 1987: 'Die Chancen der Risikogesellschaft'. *Freibeuter* 31, 149–154

Schnaiberg, A. 1980: *The Environment, from Surplus to Scarcity.* Oxford

Schnaiberg, A. et al. 1986: *Distributional Conflicts in Environmental Resource Policy.* Aldershot

Schöfbänker, G. 1988: 'Tschernobyl: Krisenmanagement und Informationspolitik in Österreich'. *Österreichische Zeitschrift für Politikwissenschaft* 1, 33–42

Scholz, M. 1986: 'Unsere Verantwortung für die Umwelt'. In Bayer. Staatsministerium für Wirtschaft und Verkehr (ed.), *Wirtschaft und Umwelt,* Reihe Tagungsberichte 1, 27–35

Schreiber, H. 1985: 'Der Preis des Wachstums – oder: Probleme der Umweltpolitik in der Volksrepublik Polen'. *Zeitschrift für Umweltpolitik und Umweltrecht* 4, 289–322

Schultze, R.-O. 1987: 'Die Bundestagswahl 1987: Eine Bestätigung des Wandels'. *Aus Politik und Zeitgeschichte,* Supplement to 'Das Parlament' 12, 3–17

Schulz, W. 1985: *Der monetäre Wert besserer Luft: Eine empirische Analyse individueller Zahlungsbereitschaften und ihrer Determinanten auf der Basis von Represäntativumfragen.* Frankfurt

Schumann, M. 1987: 'Industrielle Produzenten' in der ökologischen Herausforderung. Göttingen (research proposal)

Schumm, W. 1986: 'Die Risikoproduktion kapitalistischer Industriegesellschaften: Zur These von der Risikogesellschaft'. In R. Erd, O. Jacobi and W. Schumm (eds) 1986: *Strukturwandel in der Industriegesellschaft.* Frankfurt

Schwemmer, O. (ed.) 1987: *Über Natur: Philosophische Beiträge zum Naturverständnis.* Frankfurt

Seiferle, R. P. 1984: *Fortschrittsfeinde? Opposition gegen Technik und Industrie von der Romantik bis zur Gegenwart.* Munich

Seitz, A. 1988: *Sozialwissenschaftliche Argumentation zur Einschätzung des Bedrohungspotentials von Kernenergie*. Bamberg (Diploma thesis)

Simmel, G. 1968: *Soziologie: Untersuchungen über die Formen der Vergesellschaftung*. Berlin

Smith, J. M. 1970: 'Eugenik und Utopie'. In F. E. Mannel (ed.), *Wunschtraum und Experiment*, Freiburg, 175–97

Spaemann, R. 1984: 'Technische Eingriffe in die Natur als Problem der politischen Ethik'. In B. Guggenberger and C. Offe (eds): *An den Grenzen der Mehrheitsdemokratie: Politik und Soziologie der Mehrheitsregel*, Opladen, 240–53

SPD 1985: *Ökologische Modernisierung der Wirtschaft*. Bonn

Spelsberg, G. 1984: *Rauchplage: Hundert Jahre Saurer Regen*. Aachen

Spiegelberg, F. 1984: *Reinhaltung der Luft im Wandel der Zeit*. Düsseldorf

Sprenger, R. U. 1986: 'Umwelttechnik: Rolle und Bedeutung für Wirtschaft und Arbeitsmarkt'. In Bayer. Staatsministerium für Wirtschaft und Umwelt, Reihe Tagungsberichte 1, 97–104

Stegmüller, W. 1970: *Probleme und Resultate der Wissenschaftstheorie*. Berlin and New York

Stephens, S. 1987: 'Chernobyl Fallout: A Hard Rain for the Sami'. *Cultural Survival Quarterly* 11, 66–71

Stiller, M. 1987: 'Arsen und Spitzlasten: Deponien ängstigen die Bevölkerung'. In M. Urban (ed.), *Leben mit Chemie*, Munich, 45–50

Strasser, J. and Traube, K. 1984: *Die Zukunft des Fortschritts: Der Sozialismus und die Krise des Industrialismus*. Berlin

Tanner, J. 1988: 'Die Chemiekatastrophe von "Schweitzerhalle" und ihr Widerhall in der Schweizerischen Umweltpolitik'. *Österreichische Zeitschrift für Politikwissenschaft* 1, 17–32

Testart, J. 1988: 'Retortenkinder'. *Natur* 4, 65–6

Thompson, W. I. 1985: *Die pazifische Herausforderung: Re-Vision des politischen Denkens*. Munich

de Tocqueville, A. 1969/1975: *Democracy in America*, translated by George Lawrence. New York

Toffler, A. 1980: *The Third Wave*. London

Touraine, A. 1983: 'Soziale Bewegungen'. *Soziale Welt* 2, 143–52

Touraine, A. 1985: 'Klassen, soziale Bewegungen und soziale Schichtungen in einer nachindustriellen Gesellschaft'. In H. Strasser and J. H. Goldthorpe (eds), *Die Analyse sozialer Ungleichheit*, Opladen, 324–38

Touraine, A., Hegedus, Z., Dubet, F. and Wieviorka, M. 1982: *Die antinukleare Prophetie. Zukunftsentwürfe einer sozialen Bewegung*. Frankfurt

Treiber, H. 1983: 'Regulative Politik in der Krise?' *Kriminalsoziologische Bibliographie* 10, 28–53

Treiber, H. 1986: 'Vollzugskosten des Rechtsstaates'. *Recht und Politik* 22, 20–31

Trepl, L. 1983: 'Ökologie – eine grüne Leitwissenschaft? Über Grenzen und Perspektiven einer modernen Disziplin'. In Kursbuch 74, Berlin

Trepl, L. 1987: *Geschichte der Ökologie: Vom 17. Jahrhundert bis zur*

Gegenwart: Zehn Vorlesungen. Frankfurt

Tsuru, S. and Weidner, H. 1985: *Ein Modell für uns: Die Erfolge der japanischen Umweltpolitik.* Cologne

Umweltbundesamt 1986: *Daten zur Umwelt 1985.* Berlin

Urban, M. 1987: 'Grenzen der Belastbarkeit von Wasser, Luft und Erde'. In *idem* (ed.), *Leben mit Chemie*, Munich, 11–17

Vico, G. 1984: *The New Science of Giambattista Vico.* ed. Bergin and Fisch, 1984. Ithaca

Vig, N. J. and Kraft, M. E. 1984: *Environmental Policies in the 1980s: Reagan's New Agenda.* Washington

Vogel, D. 1986: *National Styles of Regulation: Environmental Policy in Great Britain and the United States.* London

Voigt, R. (ed.) 1986: *Recht als Instrument der Politik.* Opladen

Voltaire, F. M. 1981: *Candide.* New York

Wagner, F. 1964: *Die Wissenschaft und die gefährdete Welt.* Munich

Wagner, H., Ziegler, E. and Closs, K.-D. 1982: *Risikoaspekte der nuklearen Entsorgung.* Baden-Baden

Wambach, M. M. (ed.) 1983: *Der Mensch als Risiko: Zur Logik der Prävention und Früherkennung.* Frankfurt

Wassermann, O. 1984: 'Stellungnahme zur öffentlichen Anhörung "Auswirkungen der Luftverschmutzung auf die menschliche Gesundheit"'. Document 10/33 of the German Bundestag. Bonn

Weber, M. 1968: 'Die "Objektivität" sozialwissenschaftlicher und sozialpolitischer Erkenntnis'. In *idem, Gesammelte Aufsätze zur Wissenschaftslehre.* Tübingen

Weber, M. 1976: *The Protestant Ethic and the Spirit of Capitalism.* London

Weber, M. 1979: *Economy and Society.* Berkeley

Weber, W. 1986: *Technik und Sicherheit in der deutschen Industriegesellschaft 1850–1930* (Gesellschaft für Sicherheitswissenschaft 10). Wuppertal

Wehling, P. 1987: *Ökologische Orientierungen in der Soziologie: Sozialökologische Arbeitspapiere.* Frankfurt (duplicated manuscript)

Weidner, H. 1982: *Umweltpolitik in Japan: Erfolge und Versäumnisse.* Berlin, (manuscript; IIUG dp 82–4)

Weidner, H. 1984: 'Erfolge und Grenzen technokratischer Umweltpolitik in Japan'. *Aus Politik und Zeitgeschichte*, Supplement to *Das Parlament* 9–10, 31–46

Weidner, H. 1985: 'Umweltpolitik in Japan: Erfahrungen mit Kompensationssystemen, Abgabenregelungen und Vereinbarungen'. In R.-U. Sprenger (ed.) 1985: *Mehr Umweltschutz für weniger Geld* (IFO-Studien zur Umweltökonomie 4), Munich, 389–400

Weidner, H. and Knoepfel, P. 1985a: 'Chancen und Restriktionen kommunaler Luftreinhaltepolitik: Ein Problemaufriss'. In *idem, Luftreinhaltepolitik in städtischen Ballungsräumen: Internationale Erfahrungen*, Frankfurt, 11–34

Weidner, H. and Knoepfel, P. 1985b: 'Probleme der EG-Umweltpolitik am Beispiel der SO_2-Richtlinie'. In *idem, Luftreinhaltepolitik in städtischen*

Ballungsräumen: Internationale Erfahrungen, Frankfurt, 355–70
Weinberger, M.-L. 1987: 'Von der Müsli-Kultur zur Yuppie-Kultur'. *Die Neue Gesellschaft* 4 (Frankfurter Hefte), 352–8
Weingart, P. 1979: 'Das "Harrisburg-Syndrom" oder die De-Professionalisierung der Experten'. In H. Nowotny, *Kernenergie: Gefahr oder Notwendigkeit*, Frankfurt
Weingart, P. 1983: 'Verwissenschaftlichung der Gesellschaft: Politisierung der Wissenschaft'. *ZfS* 3, 225ff
Weingart, P. 1984: 'Eugenic Utopias'. In E. Mendelsohn and H. Nowotny (eds), *Nineteen Eighty-Four: Science between Utopia and Dystopia*, Dordrecht, 173–87
Weish, P. and Gruber, E. 1986: *Radioaktivität und Umwelt*. Stuttgart
Welsch, W. 1987: *Unsere postmoderne Moderne*. Freiburg
Wey, K.-G. 1982: *Umweltpolitik in Deutschland: Kurze Geschichte des Umweltschutzes in Deutschland seit 1900*. Opladen
Wieseltier, L. 1983: *Nuclear War, Nuclear Peace*. Washington
Winter, G. 1986: 'Gentechnik als Rechtsproblem'. *Deutsches Verwaltungsblatt* 12, 585–96
Winter, G. 1987: 'Die Angst des Richters vor dem Recht: Über gerichtliche Massstäbe der Technikkontrolle'. In B. Lutz (ed.), *Technik und sozialer Wandel: Verhandlungen des 23. Deutschen Soziologentages in Hamburg 1986*, Frankfurt, 464–71
Winter, G. and Schäfer, R. 1985: 'Zur richterlichen Rezeption natur- und ingenieurwissenschaftlicher Voraussagen über komplexe technische Systeme am Beispiel von Kernkraftwerken'. *NVwZ* 10, 703–11
Wörle, R. 1986: 'Technischer Umweltschutz als Fach- und Vollzugsaufgabe der Verwaltung'. In Bayer. Staatsministerium für Wirtschaft und Verkehr (ed.), *Wirtschaft und Umwelt*, Reihe Tagungsberichte 1, 87–96
von Woldeck, R. 1986: 'Warnung vor der Wahrscheinlichkeit'. In Kursbuch 85: *GAU: Die Havarie der Expertenkultur*, Berlin, 63–80
Wolf, R. 1986a: 'Das Recht im Schatten der Technik'. *Kritische Justiz* 19, 241–62
Wolf, R. 1986b: *Der Stand der Technik: Geschichte, Strukturelemente und Funktion der Verechtlichung technischer Risiken am Beispiel des Immissionschutzes*. Opladen
Wolf, R. 1987: 'Zur Antiquiertheit des Rechts in der Risikogesellschaft'. *Leviathan* 15, 357–91
Wolf, R. 1988: '"Herrschaft kraft Wissen" in der Risikogesellschaft'. *Soziale Welt* 2, 164–87
Wollschläger, H. 1987: 'Tiere sehen Dich an oder das Potential Mengele'. *Die Republik* (ed. P. and U. Nettelbeck), 79–81 (April)
Wynne, B. 1983: '"Technologie, Risiko und Partizipation: Zum gesellschaftlichen Umgang mit Unsicherheit"'. In J. Conrad (ed.), *Gesellschaft, Technik und Risikopolitik*, Berlin
Zellentin, G. 1979: *Abschied vom Leviathan: Ökologische Aufklärung über*

politische Alternativen. Hamburg

Ziegler, C. E. 1986a: *Environmental Protection in the Soviet Union.* Berlin (manuscript; IIUG pre 86–17)

Ziegler, C. E. 1986b: 'Issue Creation and Interest Groups in Soviet Environmental Policy: The Applicability of the State Corporatist Model'. *Comparative Politics* 18, 171–92

Ziegler, E. 1987: 'Das deutsche Atomrecht vor und nach Tschernobyl: Neuere Entwicklungen im Atom- und Strahlenschutzrecht'. *Energiewirtschaftliche Tagesfragen* 37, 353–60

Zimmerli, W. 1985: 'Dynamik der Wissenschaftsentwicklung und Wandel fundamentaler Werte'. *Zeitschrift für Wissenschaftsforschung* (December), 22–34

Zimmerli, W. 1986: 'Ausstieg aus der Ethik?' *Der Spiegel* 24, 45–6

Zimmermann, K. 1985: *'Präventive' Umweltpolitik und technologische Anpassung.* Berlin (manuscript; IIUG dp 85–8)

Index